Electrochemistry in light water reactors

European Federation of Corrosion Publications
NUMBER 49

Electrochemistry in light water reactors

Reference electrodes, measurement, corrosion and tribocorrosion issues

---

Edited by
R.-W. Bosch, D. Féron and J.-P. Celis

Published for the European Federation of Corrosion
by Woodhead Publishing and Maney Publishing
on behalf of
The Institute of Materials, Minerals & Mining

CRC Press
Boca Raton Boston New York Washington, DC

WOODHEAD PUBLISHING LIMITED
Cambridge England

Woodhead Publishing Limited and Maney Publishing Limited on behalf of
The Institute of Materials, Minerals & Mining

Woodhead Publishing Limited, Abington Hall, Abington,
Cambridge CB21 6AH, England
www.woodheadpublishing.com

Published in North America by CRC Press LLC, 6000 Broken Sound Parkway, NW,
Suite 300, Boca Raton, FL 33487, USA

First published 2007 by Woodhead Publishing Limited and CRC Press LLC

British Library Cataloguing in Publication Data
A catalogue record for this book is available from the British Library.

Library of Congress Cataloging in Publication Data
A catalog record for this book is available from the Library of Congress.

Woodhead Publishing ISBN 978-1-84569-240-7 (book)
Woodhead Publishing ISBN 978-1-84569-302-2 (e-book)
CRC Press ISBN 978-1-4200-5408-8
CRC Press order number WP5408
ISSN 1354-5116

Typeset by SNP Best-set Typesetter Ltd., Hong Kong
Printed by T J International Limited, Padstow, Cornwall, England

Contents

Contributor contact details

(* = main contact)

Editors

R.-W. Bosch*
SCK·CEN
Belgian Nuclear Research Centre
Reactor Materials Research Unit
Boeretang 200
2400 Mol
Belgium
E-mail: rbosch@sckcen.be

D. Féron
CEA – Saclay
DPC/SCCME
Bât. 458, P.C. 50
91191 Gif-sur-Yvette cédex
France
E-mail: damien.feron@cea.fr

J.-P. Celis
KU Leuven
Dept MTM
Kasteelpark Arenberg 44
BE-3001 Leuven
Belgium
E-mail: jcan-pierre.celis@kuleuven.be

Chapter 1

Digby D. Macdonald*, Ting Zhu and Xueyong Guan
Center for Electrochemical Science and Technology
Department of Materials Science and Engineering
Pennsylvania State University
201 Steidle Bldg
University Park
PA 16802
USA
E-mail: ddm2@psu.edu

Chapter 2

R.-W. Bosch*
SCK·CEN
Belgian Nuclear Research Centre
Reactor Materials Research Unit
Boeretang 200
2400 Mol
Belgium
E-mail: rbosch@sckcen.be

Z. Kerner, G. Nagy
AEKI
Konkoly-Thege út 29–33
Budapest
Hungary
E-mail: nagyg@sunserv.kfki.hu

D. Féron
CEA – Saclay
DPC/SCCME
Bât. 458, P.C. 50
91191 Gif-sur-Yvette cédex
France
E-mail: damien.feron@cea.fr

M. Navas
CIEMAT
Technology Department
Avenida Complutense 22
Madrid 28040
Spain
E-mail: m.navas@ciemat.es

W. Bogaerts
KU Leuven
Dept MTM
Kasteelpark Arenberg 44
BE-3001 Leuven
Belgium
E-mail: walter.bogaerts@mtm.kuleuven.be

D. Kárník
NRI Rez
Rez 130
Rez
Czech Republic
E-mail: kar@nri.cz

T. Dorsch, R. Kilian
Framatone ANP
Freyeslebenstrasse 1
Erlangen
Germany
E-mail: renate.kilian@framatome-anp.de

M. Ullberg, A. Molander
Studsvik Nuclear AB
SE-611 82 Nyköping
Sweden
E-mail: anders.molander@studsvik.se

K. Mäkelä, T. Saario
VTT Industrial Systems
Kemistintie 3
Espoo
Finland
E-mail: kari.makela@vtt.fi

Chapter 3

Anders Molander
Studsvik Nuclear AB
SE-611 82 Nyköping
Sweden
E-mail: anders.molander@studsvik.se

Chapter 4

Marta Navas*, Marisol García
CIEMAT
Technology Department
Avenida Complutense 22
Madrid 28040
Spain
E-mail: m.navas@ciemat.es

Chapter 5

M. Postler*
NRI Řež
Husinec-Řež, 250 68
Czech Republic
E-mail: pom@ujv.cz

R.-W. Bosch, S. Van Dyck
SCK·CEN
Belgian Nuclear Research Centre
Reactor Materials Research Unit
Boeretang 200
2400 Mol
Belgium
E-mail: rbosch@sckcen.be

Chapter 6

Gabor Nagy*, Zsolt Kerner, János Balog, Robert Schiller
KFKI Atomic Energy Research Institute
H-1525 Budapest
P.O. Box 49
Hungary
E-mail: nagyg@sunserv.kfki.hu

Chapter 7

M. Tupin*, C. Bataillon, J. P. Gozlan, P. Bossis
Laboratoire de Microscopie et d'Etude de l'Endommagement
DMN/SEMI/LM2E
CEA – Saclay
91191 Gif-sur-Yvette cédex
France
E-mail: marc.tupin@cea.fr

Chapter 8

R. Burrows*, S. Harris
Nexia Solutions Ltd
Berkeley Centre
Berkeley
Gloucestershire
GL13 9PB
UK
E-mail: robert.burrows@nexiasolutions.com

Chapter 9

J-Ph. Vernot*, D. Kaczorowski
AREVA NP – Centre Technique
Départment Mécanique des fluides – Mécanique des Structures (NTT-F)
Section Mécanique Statique, Usure Frottement et Fatigue
Porte Magenta
1 rue B. Marcet
BP 181
71205 Le Creusot cédex
France
E-mail: jeanphilippe.vernot@areva.com

Chapter 10

D. Déforge*, A. Ambard, A. Lina, P. Ponthiaux, F. Wenger
Electricité de France (EDF) R&D
Chemistry and Corrosion Group
Les Renardières
Avenue des Renardières
Ecuelles F-77818
Moret-sur-Loing cédex
France
E-mail: antoine.ambard@edf.fr
wenger@lgpm.ecp.fr

Chapter 11

François Wenger*, Pierre Ponthiaux, Lidia Benea, Jean Peybernès, Antoine Ambard
Ecole Centrale Paris
Laboratoire LGPM
Grande Voie des Vignes
92295 Châtenay-Malabry
France
E-mail: wenger@lgpm.ecp.fr

European Federation of Corrosion (EFC) publications: Series introduction

The EFC, incorporated in Belgium, was founded in 1955 with the purpose of promoting European co-operation in the fields of research into corrosion and corrosion prevention.

Membership of the EFC is based upon participation by corrosion societies and committees in technical Working Parties. Member societies appoint delegates to Working Parties, whose membership is expanded by personal corresponding membership.

The activities of the Working Parties cover corrosion topics associated with inhibition, education, reinforcement in concrete, microbial effects, hot gases and combustion products, environment sensitive fracture, marine environments, refineries, surface science, physico-chemical methods of measurement, the nuclear industry, the automotive industry, computer-based information systems, coatings, tribocorrosion and the oil and gas industry. Working Parties and Task Forces on other topics are established as required.

The Working Parties function in various ways, e.g. by preparing reports, organising symposia, conducting intensive courses and producing instructional material, including films. The activities of the Working Parties are co-ordinated, through a Science and Technology Advisory Committee, by the Scientific Secretary. The administration of the EFC is handled by three Secretariats: DECHEMA e.V. in Germany, the Société de Chimie Industrielle in France, and The Institute of Materials, Minerals and Mining in the United Kingdom. These three Secretariats meet at the Board of Administrators of the EFC. There is an annual General Assembly at which delegates from all member societies meet to determine and approve EFC policy. News of EFC activities, forthcoming conferences, courses, etc., is published in a range of accredited corrosion and certain other journals throughout Europe. More detailed descriptions of activities are given in a Newsletter prepared by the Scientific Secretary.

The output of the EFC takes various forms. Papers on particular topics, for example, reviews or results of experimental work, may be published in scientific and technical journals in one or more countries in Europe. Conference proceedings are often published by the organisation responsible for the conference.

In 1987 the, then, Institute of Metals was appointed as the official EFC publisher. Although the arrangement is non-exclusive and other routes for publication are still available, it is expected that the Working Parties of the EFC will use The Institute of Materials, Minerals and Mining for publication of reports, proceedings, etc., wherever possible.

The name of The Institute of Metals was changed to The Institute of Materials on 1 January 1992 and to The Institute of Materials, Minerals and Mining with effect from 26 June 2002. The series is now published by Woodhead Publishing and Maney Publishing on behalf of The Institute of Materials, Minerals and Mining.

P. McIntyre
EFC Series Editor, The Institute of Materials, Minerals and Mining, London, UK EFC

Secretariats are located at:

Dr B. A. Rickinson
European Federation of Corrosion, The Institute of Materials, Minerals and Mining, 1 Carlton House Terrace, London SW1Y 5DB, UK

Dr J. P. Berge
Fédération Européenne de la Corrosion, Société de Chimie Industrielle, 28 rue Saint-Dominique, F-75007 Paris, FRANCE

Professor Dr G. Kreysa
Europäische Föderation Korrosion, DECHEMA e.V., Theodor-Heuss-Allee 25, D-60486 Frankfurt, GERMANY

Volumes in the EFC series

1 **Corrosion in the nuclear industry**
Prepared by the Working Party on Nuclear Corrosion

2 **Practical corrosion principles**
Prepared by the Working Party on Corrosion Education (out of print)

3 **General guidelines for corrosion testing of materials for marine applications**
Prepared by the Working Party on Marine Corrosion

4 **Guidelines on electrochemical corrosion measurements**
Prepared by the Working Party on Physico-Chemical Methods of Corrosion Testing

5 **Illustrated case histories of marine corrosion**
Prepared by the Working Party on Marine Corrosion

6 **Corrosion education manual**
Prepared by the Working Party on Corrosion Education

7 **Corrosion problems related to nuclear waste disposal**
Prepared by the Working Party on Nuclear Corrosion

8 **Microbial corrosion**
Prepared by the Working Party on Microbial Corrosion

9 **Microbiological degradation of materials – and methods of protection**
Prepared by the Working Party on Microbial Corrosion

10 **Marine corrosion of stainless steels: chlorination and microbial effects**
Prepared by the Working Party on Marine Corrosion

11 **Corrosion inhibitors**
Prepared by the Working Party on Inhibitors (out of print)

12 **Modifications of passive films**
Prepared by the Working Party on Surface Science and Mechanisms of Corrosion and Protection

13 **Predicting CO_2 corrosion in the oil and gas industry**
Prepared by the Working Party on Corrosion in Oil and Gas Production (out of print)

14 **Guidelines for methods of testing and research in high temperature corrosion**
Prepared by the Working Party on Corrosion by Hot Gases and Combustion Products

15 **Microbial corrosion (Proc. 3rd Int. EFC Workshop)**
Prepared by the Working Party on Microbial Corrosion

16 **Guidelines on materials requirements for carbon and low alloy steels for H_2S-containing environments in oil and gas production**
Prepared by the Working Party on Corrosion in Oil and Gas Production

17 **Corrosion resistant alloys for oil and gas production: guidance on general requirements and test methods for H_2S Service**
Prepared by the Working Party on Corrosion in Oil and Gas Production

18 **Stainless steel in concrete: state of the art report**
Prepared by the Working Party on Corrosion of Reinforcement in Concrete

19 **Sea water corrosion of stainless steels – mechanisms and experiences**
Prepared by the Working Parties on Marine Corrosion and Microbial Corrosion

20 **Organic and inorganic coatings for corrosion prevention – research and experiences**
Papers from EUROCORR '96

21 **Corrosion–deformation interactions**
CDI '96 in conjunction with EUROCORR '96

22 **Aspects on microbially induced corrosion**
Papers from EUROCORR '96 and the EFC Working Party on Microbial Corrosion

23 **CO_2 corrosion control in oil and gas production – design considerations**
Prepared by the Working Party on Corrosion in Oil and Gas Production

24 **Electrochemical rehabilitation methods for reinforced concrete structures – a state of the art report**
Prepared by the Working Party on Corrosion of Reinforced Concrete

25 **Corrosion of reinforcement in concrete – monitoring, prevention and rehabilitation**
Papers from EUROCORR '97

26 **Advances in corrosion control and materials in oil and gas production**
Papers from EUROCORR '97 and EUROCORR '98

27 **Cyclic oxidation of high temperature materials**
Proceedings of an EFC Workshop, Frankfurt/Main, 1999

28 **Electrochemical approach to selected corrosion and corrosion control**
Papers from 50th ISE Meeting, Pavia, 1999

29 **Microbial corrosion (Proc. 4th Int. EFC workshop)**
Prepared by the Working Party on Microbial Corrosion

30 **Survey of literature on crevice corrosion (1979–1998): mechanisms, test methods and results, practical experience, protective measures and monitoring**
Prepared by F. P. Ijsseling and the Working Party on Marine Corrosion

31 **Corrosion of reinforcement in concrete: corrosion mechanisms and corrosion protection**
Papers from EUROCORR '99 and the Working Party on Corrosion of Reinforcement in Concrete

32 **Guidelines for the compilation of corrosion cost data and for the calculation of the life cycle cost of corrosion – a working party report**
Prepared by the Working Party on Corrosion in Oil and Gas Production

33 **Marine corrosion of stainless steels: testing, selection, experience, protection and monitoring**
Edited by D. Féron on behalf of the Working Party on Marine Corrosion

34 **Lifetime modelling of high temperature corrosion processes**
Proceedings of an EFC Workshop 2001. Edited by M. Schütze, W. J. Quadakkers and J. R. Nicholls

35 **Corrosion inhibitors for steel in concrete**
Prepared by B. Elsener with support from a Task Group of Working Party 11 on Corrosion of Reinforcement in Concrete

36 **Prediction of long term corrosion behaviour in nuclear waste systems**
Edited by D. Féron and Digby D. Macdonald of Working Party 4 on nuclear corrosion

37 **Test methods for assessing the susceptibility of prestressing steels to hydrogen induced stress corrosion cracking**
Prepared by B. Isecke on behalf of Working Party 11 on Corrosion of Reinforcement in Concrete

38 **Corrosion of reinforcement in concrete: mechanisms, monitoring, inhibitors and rehabilitation techniques**
Edited by M. Raupach, B. Elsener, R. Polder and J. Mietz on behalf of Working Party 11 on Corrosion of Steel in Concrete

39 **The use of corrosion inhibitors in oil and gas production**
Edited by J. W. Palmer, W. Hedges and J. L. Dawson

40 **Control of corrosion in cooling waters**
Edited by J. D. Harston and F. Ropital

41 **Metal dusting, carburisation and nitridation**
Edited by M. Schütze and H. Grabke

42 **Corrosion in refineries**
Edited by J. Harston

43 **The electrochemistry and characteristics of embeddable reference electrodes for concrete**
Prepared by R. Myrdal on behalf of Working Party 11 on Corrosion of Steel in Concrete

44 **The use of electrochemical scanning tunnelling microscopy (EC-STM) in corrosion analysis: reference material and procedural guidelines**
Prepared by R. Lindström, V. Maurice, L. Klein and P. Marcus on behalf of Working Party 6 on Surface Science

45 **Local probe techniques for corrosion research**
Edited by R. Oltra (on behalf of Working Party 8 on Physico-Chemical Methods of Corrosion Testing)

46 **Amine unit corrosion survey**
Edited by F. Ropital (on behalf of Working Party 15 on Corrosion in the Refinery Industry)

47 **Novel approaches to the improvement of high temperature corrosion resistance**
Edited by M. Schütze and W. Quadakkers (on behalf of Working Party 3 on Corrosion in Hot Gases and Combustion Products)

48 **Corrosion of metallic heritage artefacts: investigation, conservation and prediction of long term behaviour**
Edited by P. Dillmann, G. Béranger, P. Piccardo and H. Matthiessen (on behalf of Working Party 4 on Nuclear Corrosion)

49 **Electrochemistry in light water reactors: reference electrodes, measurement, corrosion and tribocorrosion**
Edited by R.-W. Bosch, D. Féron and J.-P. Celis (on behalf of Working Party 4 on Nuclear Corrosion)

50 **Corrosion behaviour and protection of copper and aluminium alloys in seawater**
Edited by D. Féron (on behalf of Working Party 9 on Marine Corrosion)

51 **Corrosion issues in light water reactors: stress corrosion cracking**
Edited by D. Féron and J.-M. Olive (on behalf of Working Party 4 on Nuclear Corrosion)

Preface

Four years ago, the European Community project LIRES (Development of Light Water Reactor Reference Electrodes) was started with the objective of developing high temperature reference electrodes for light water reactors (LWRs). To conclude this research project, a one-day seminar was organized with the topic: 'High Temperature Electrochemistry in Light Water Reactors' (EFC event No. 279, 30 November 2004, KU Leuven, Belgium). The objective of the seminar was to give an overview of the developments of electrochemical methods and techniques for use in high temperature, high pressure aqueous environments in Europe and to focus on their application to LWRs. This volume publishes the fruits of all the technical and scientific discussions which occurred during the friendly one-day seminar, including a selection of papers presented during this seminar and papers presented during EUROCORR conferences. These papers have been brought together as they all concern corrosion problems in LWRs, attempting to solve them using electrochemical methods.

The volume has been divided into three parts: (1) Measurements and reference electrodes, (2) Electrochemistry and corrosion issues, and (3) Electrochemistry and tribocorrosion issues. The current state-of-the-art is described, showing the main difference between the electrochemistry in boiling and in pressurised water reactors (PWRs). The volume also covers topics ranging from reference electrode developments, electrochemical tests, in-pile testing and tribocorrosion behaviour of stainless steel under PWR conditions.

We would like to thank the authors who presented and wrote papers of outstanding scientific and technical content and who responded enthusiastically to the questions and comments raised by the reviewers; the members of the EFC Working Party 4 on 'nuclear corrosion' (WP4) and EFC Working Party 18 on 'tribocorrosion' (WP18) who reviewed and commented on these papers; and also Miss Mylene Belgome, Secretary of WP4, who helped the editors in the secretarial procedures reviewing.

We hope that the reader will enjoy the papers and realize that even under severe conditions (e.g. autoclave testing at high temperatures and pressures, in-reactor testing) electrochemical methods give valuable results, and that this volume will be useful to scientists and engineers for the understanding and resolution of corrosion phenomena in LWRs.

Rik-Wouter Bosch, Damien Féron and Jean-Pierre Celis

Part I

Measurements and reference electrodes

1

Current state-of-the-art in reference electrode technology for use in high subcritical and supercritical aqueous systems

DIGBY D. MACDONALD, TING ZHU and XUEYONG GUAN, Pennsylvania State University, USA

1.1 Introduction

Aqueous systems under the conditions of high temperature and high pressure are receiving ever-increasing attention, because of the use of water as a heat transfer fluid in thermal (electrical) power generation and as a reaction medium in supercritical water oxidation (SCWO) technology, and because of their existence or involvement in deep ocean hydrothermal vents, hydrothermal ore-body formation processes, deep natural wells, fluid sedimentary basins, nuclear waste disposal, geothermal power plants, hydrothermal crystal growth, etc. All of these systems involve aqueous solutions at temperatures ranging up to about 360°C (high subcritical) to 650°C (supercritical systems), with both regimes commonly occurring in a single system. As shown below, the chemical, physical and thermodynamic properties of aqueous solutions undergo profound changes over these temperature ranges, and these changes are reflected in their interactions with metals and alloys. These interactions are primarily electrochemical in nature, resulting in corrosion, for example. The ability to monitor the chemical and electrochemical properties of high subcritical and supercritical aqueous systems is essential to the development of many of the pending technological applications, such as advanced power generation via supercritical nuclear reactors (SCNRs). pH is one of the most important chemical parameters defining the reaction medium and corrosion properties of high subcritical and of supercritical aqueous environments and viable methods for measuring pH continue to be developed. In particular, high quality pH and potentiometric measurements are amongst the most powerful tools available for exploring the thermodynamics and kinetics of reactions that occur in high temperature aqueous solutions.

Aqueous systems at elevated temperatures and pressures possess many unusual properties, especially when the temperature and the pressure exceed the critical values ($T = 374.15$°C, $p = 221$ bar) [1]. While space does not permit a detailed review of these properties here, and because a number of excellent reviews are available in the literature [2], only some of the more relevant properties are discussed below. The properties that are most relevant to reference electrode technology are as follows:

1 The density of supercritical water is approximately 0.1 g cm^{-3} at 500°C and 300 bar (Fig. 1.1 [3]), which is comparable to the density of nitrogen gas at 25°C and 100 bar. Accordingly, supercritical water under these typical chemical process conditions (e.g., SCWO) (400–700°C, 100–700 bar) is better described as a 'dense gas' rather than a 'liquid'. (Thermodynamically, supercritical fluids are considered to be a state different from the solid, liquid and gaseous state. Because water has only one phase under supercritical conditions, according to Gibbs' phase rule, there are two degrees of freedom for the system, thus the pressure and temperature may be varied independently while maintaining the same single phase.)
2 With increasing temperature at constant pressure, again under typical process conditions, the dielectric constant decreases rapidly to a value that is typical of a nonpolar solvent (Fig. 1.2 [3]). Consequently, strong

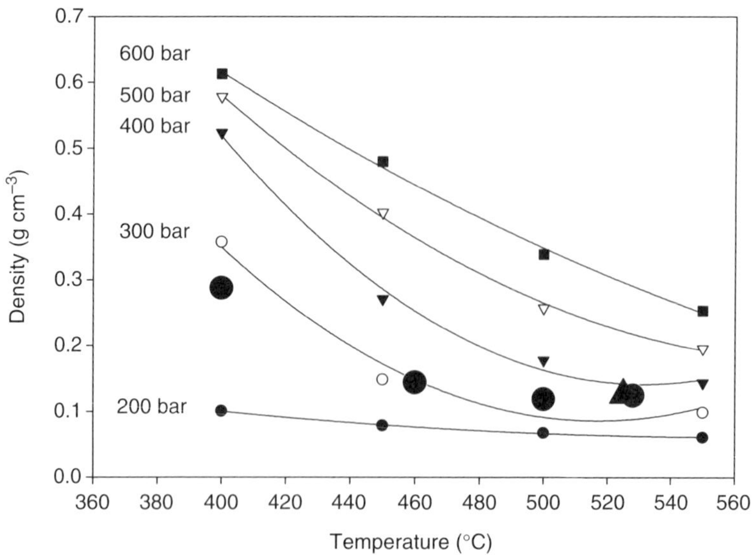

Fig. 1.1 Density versus temperature for supercritical water as a function of pressure [3]. The large filled circles represent the conditions employed in Ref. 3 to evaluate solution (0.01 m HCl) acidity. The large filled triangle corresponds to the conditions that existed in the thermohydrolysis of carbon tetrachloride (see Fig. 1.22b).

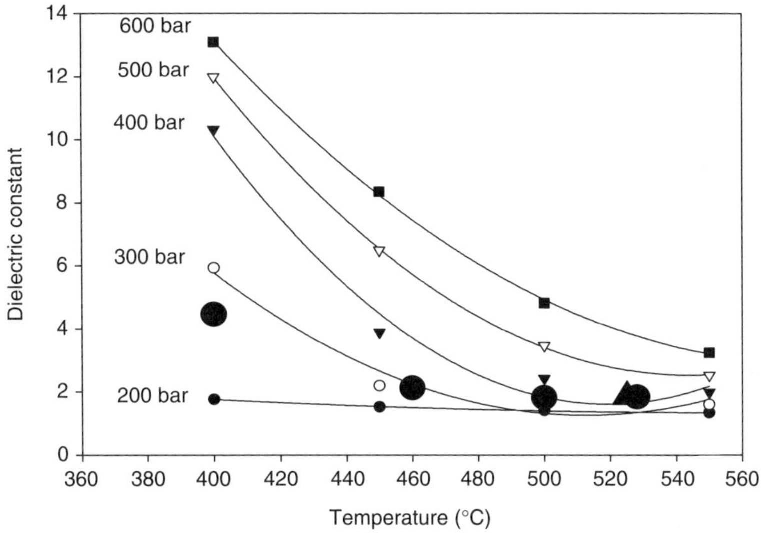

Fig. 1.2 Dielectric constant vs temperature for supercritical water as a function of pressure [3]. The large filled circles designate the conditions at which potentiometric acidity measurements have been reported in Ref. 3. The large filled triangle corresponds to the conditions that existed in the thermohydrolysis of carbon tetrachloride (see Fig. 1.22b).

electrolytes (e.g., HCl, NaCl and NaOH) under ambient conditions are very poorly ionized in supercritical water systems, particularly at low pressure, and hence are generally insoluble. On the other hand, many organic substances that display almost no solubility in water under ambient conditions (e.g., benzene) are fully miscible with water at supercritical temperatures. At constant temperature, the dielectric constant increases with increasing pressure (and density), particularly at low supercritical temperatures (Fig. 1.3 [3]), but the dielectric constant does not approach that of water at ambient temperature, even at extreme pressures.

3 The viscosity of supercritical water decreases as the temperature increases (Fig. 1.4 [3]). According to the inverse relationship between viscosity and diffusivity ($D \cdot \eta$ = constant), the diffusivity of species and mass transport-controlled reaction rates are expected to increase significantly as the temperature increases at constant pressure.

4 From room temperature to 255°C, thc value of the ion product of water (K_w) monotonically increases to a maximum of about three orders of magnitude higher compared with the ambient temperature value, and then decreases slowly as the temperature is raised to 350°C. The value of K_w then begins to decrease sharply as the temperature transitions the

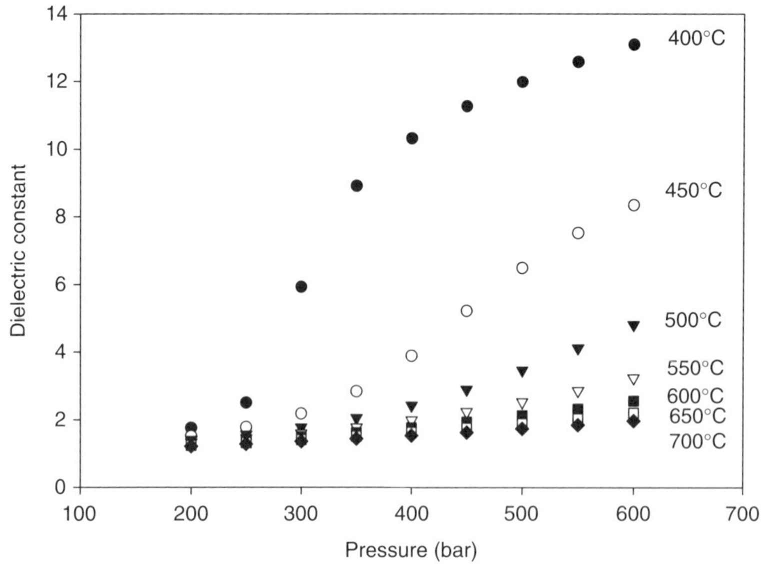

Fig. 1.3 Variation of the dielectric constant of water over extended ranges in temperature and pressure, with pressure being the primary independent variable [3].

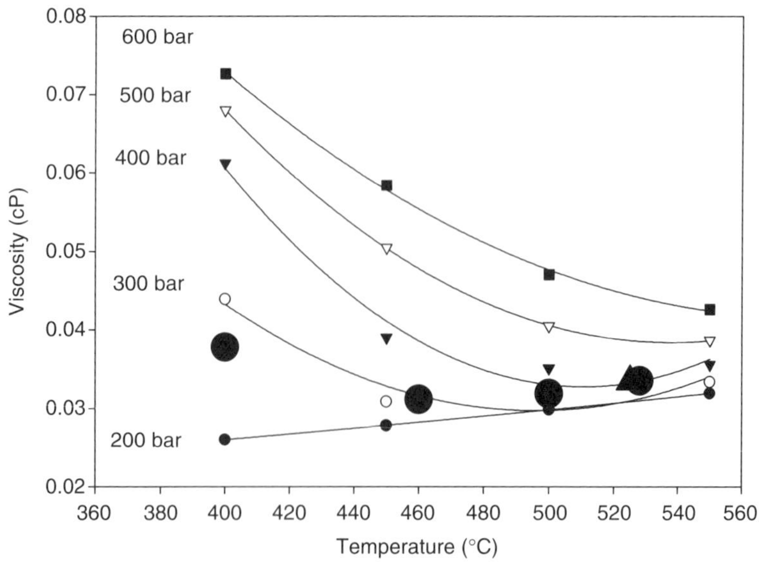

Fig. 1.4 Plots of viscosity vs temperature for supercritical water as a function of pressure [3]. The large filled circles represent the conditions under which acidity studies were performed (see Ref. 3). The large filled triangle corresponds to the conditions that existed in the thermohydrolysis of carbon tetrachloride (see Fig. 1.22b).

critical value, such that the value is reduced by nine orders of magnitude at 550°C and 250 bar compared to its value at ambient temperature and pressure. This sharp decrease in K_w provides further evidence of the nonpolar nature of supercritical water [4].

Because of the nonpolar properties of supercritical water, salts display low solubility and are readily precipitated as solid phases when supercritical conditions are achieved; this is in contrast with the majority of gas–water and water–organic systems, in which the components are fully miscible. Depending on whether critical behavior is observed in saturated solutions, salt–water systems are consequently classified into two types [5]. Type 1 systems exhibit high salt solubility in the vicinity of the critical temperature of water, whereas Type 2 systems have low salt solubility. Classical examples of Type 1 and 2 are the $NaCl–H_2O$ and $Na_2SO_4–H_2O$ systems, respectively (Fig. 1.5 [4] and 1.6 [4]). There are several interesting features in Fig. 1.5: two lines of S–L–V (solid–liquid–vapor) equilibrium (solidus) at 450 and 720°C between which there is no liquid phase present, the critical curves separating the liquid and liquid + vapor phase (L/V + L), and the saturation lines separating the liquid and liquid + solid phase (L/L + S). It is more convenient to select several concentrations to explain the phase changes that occur with increasing temperature. For the 20% (weight percentage) NaCl–H_2O system, starting from the lowest temperature shown in the diagram, the system maintains a single phase (an unsaturated NaCl solution) until it crosses the critical curve where it becomes a two-phase system:

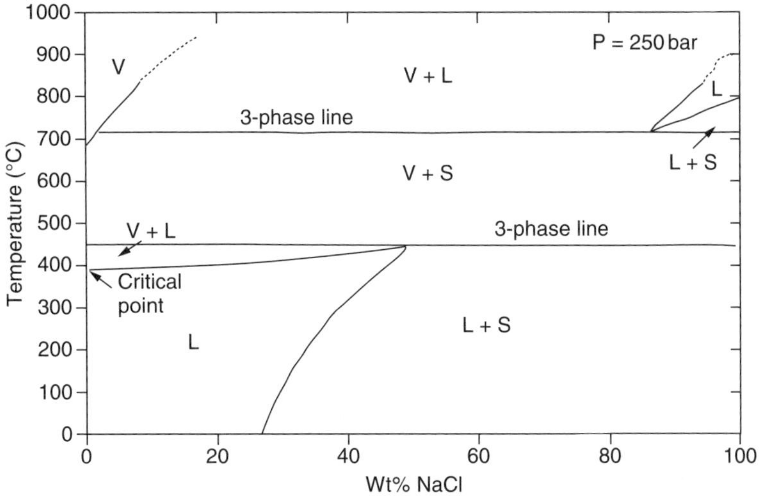

Fig. 1.5 Temperature–composition phase diagram for $NaCl–H_2O$ at 250 bar [4]. S, solid; L, liquid; V, vapor.

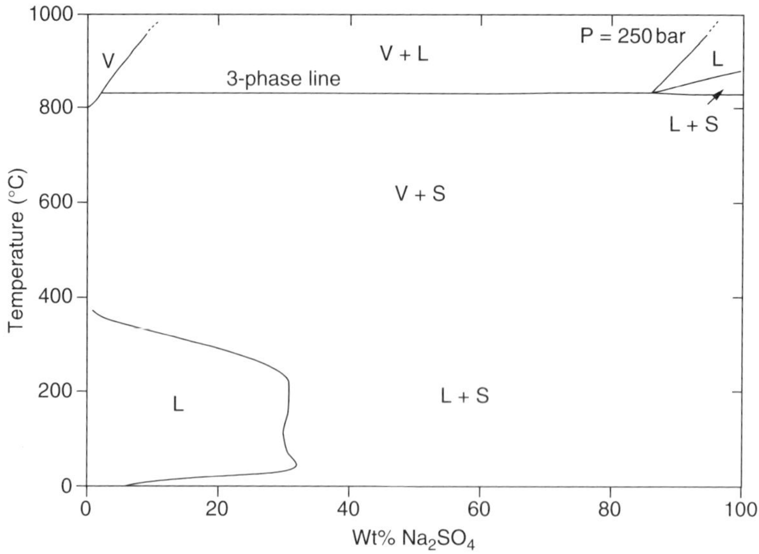

Fig. 1.6 Temperature–composition phase diagram for Na_2SO_4–H_2O at 250 bar [4]. S, solid; L, liquid; V, vapor.

a vapor phase (supercritical fluid) and a liquid phase (NaCl-saturated solution). After that, as the temperature is increased further, the system moves along the critical curve, with the NaCl concentration in the liquid progressively increasing, until the critical temperature of the saturated liquid is reached. At this point (450°C and ≈48 wt% NaCl), a 'eutectic-like' reaction occurs in the system with the formation of a 'dry salt' and supercritical water containing a very low concentration of NaCl. For the 40 wt% NaCl case, the system starts as a saturated solution with precipitated solid salt and, as the temperature increases, the NaCl solubility in the liquid solution also increases until the saturation line is met at about 300°C. At higher temperatures, the liquid phase is unsaturated and no solid salt exists in the system. However, at a still higher temperature the system intersects the critical line resulting in two phases: a supercritical water ('vapor') phase and a salt solution, as described above.

The concentration of the salt solution increases with further increases in temperature until the critical curve intersects the three-phase line at 450°C, at which point solid NaCl precipitates. Starting at 80 wt%, the NaCl system is basically the same as the previous case, except that the solution at subcritical temperatures never becomes unsaturated and a solid NaCl phase always exists in the system. At the temperature of 720°C, another eutectic reaction occurs whence the system produces a supercritical water phase containing a finite concentration of NaCl and a liquid comprising molten salt containing a certain amount of H_2O. Note that water depresses the melting temperature of NaCl. For the Na_2SO_4–H_2O system, one obvious distinction is the lack of low temperature intersection of the L + S saturation line with the S–L–V curve in the phase diagram; instead the saturated fluid displays a continuous transition from liquid-like to gas-like behavior as the temperature increases. Note also the two turning points on the saturation line at the lower part of the diagram; the solubility of Na_2SO_4 begins a slow decrease at the first point ($T \approx 40°C$) and decreases dramatically above the second point ($T \approx 200°C$) until $T = 374.5°C$, where the system is separated into two phases: supercritical fluid and solid salt.

Recent advances in the development of several types of high temperature ceramic membrane and metal–metal oxide pH sensors [e.g., the yttria-stabilized zirconia membrane electrodes (YSZMEs [6]) and the W/WO_3 sensor] offer an opportunity for accurate potentiometric studies and pH measurements in supercritical aqueous solutions at temperatures up to and above 500°C. However, these measurements are possible only with the development of correspondingly accurate reference electrodes. A generally applicable reference electrode for use in high subcritical and in supercritical aqueous environments has not emerged from the more than three decades of work in this field and, in the opinion of the authors, the development of such an electrode is the most pressing need in the field. This paper reviews the performance of a number of reference electrodes used in high subcritial and in supercritical environments and discusses some of the challenges that must be addressed in moving reference electrode technology forward.

The general requirements for a viable reference electrode that is capable of operating at temperatures up to 600°C have been described by Macdonald and Kriksunov [7]. Because of the harsh working conditions that frequently exist in high subcritical and in supercritical aqueous environments, several prerequisites must be met for the development of a viable high temperature reference electrode technology:

1 High resistance of structural components to chemical degradation, primarily to avoid contamination of the reference electrode internal solution.
2 Use of an electrochemically active element that is stable against thermal hydrolysis and/or chemical reduction/oxidation under the prevailing conditions.
3 Suppression of irreversible processes that give rise to nonequilibrium contributions to the measured potential, such as the isothermal liquid junction potential (ILJP) and the thermal liquid junction potential (TLJP).

4 Equilibrium operation, such that the measured potential of an indicator electrode can be placed on the standard hydrogen electrode (SHE) scale.
5 Accurate resolution of the measured potentials to within a few millivolts.

Two approaches have been employed in the past for potential monitoring at high temperatures: (i) the use of an internal reference electrode, in which the electroactive element operates within the high temperature environment, and (ii) the use of an external pressure balanced reference electrode, in which the electroactive element is maintained at ambient or near ambient temperature. The latter approach avoids the thermal hydrolysis of the electroactive element, which is a major cause of drift in the potential of internal reference electrodes, but it incurs an almost equally troublesome thermal liquid junction potential.

This review may be regarded as the follow-on to the review of high temperature reference electrodes published by one of the authors (D.D.M.) in 1978 [7]. That review covered only subcritical systems at temperatures up to 300°C, because little interest existed at that time in the electrochemical properties of aqueous systems at higher temperatures. With that interest now firmly established, due to the development of supercritical thermal power plants, SCWO destruction of resilient organic waste, supercritical synthesis, and exploration of deep ocean vents, to name but a few reasons, the present review covers high subcritical (300–374.15°C) and supercritical ($T > 374.15$°C) aqueous systems.

1.2 Internal reference electrodes

1.2.1 Hydrogen electrodes

From the theoretical point of view, the hydrogen electrode is perhaps the most ideal reference electrode for high temperature work. Thus, the SHE has been selected in thermodynamics as the zero of the potential scale and it is the electrode against which the standard potentials of all other half-cells are expressed. However, it is important to note that the SHE is a hypothetical construct, in which the activity of H^+ (a_{H^+}) and the fugacity of H_2 (f_{H_2}) are equal to one at the specified temperature and pressure. Although it is possible to devise a practical hydrogen electrode that conforms closely to the SHE, by choosing the appropriate concentration of H^+ and pressure of H_2, such that the activity and fugacity, respectively, are unity, this is seldom if ever done, because of the ease of relating the potential of a practical hydrogen electrode to the SHE scale (see below) and because of the difficulty in interfacing a gas (hydrogen of precisely controlled pressure) with a high temperature/high pressure aqueous solution. Even so, with appropriate care, the hydrogen electrode has proven to be rugged and reversible, but it can only be used in systems that are stable against reduction by hydrogen. For example, it cannot be used in potentiometric studies of Fe^{3+} hydrolysis, because of the occurrence of the reaction $Fe^{3+} + \frac{1}{2}H_2 \rightarrow Fe^{2+} + H^+$.

According to the 1960 Stockholm Convention, the half-cell reaction for the hydrogen electrode is expressed as:

$$2H^+ + 2e^- \Leftrightarrow H_2 \quad [1.1]$$

with the potential being given by the Nernst equation:

$$E_{H^+/H_2} = -\frac{2.303RT}{2F}\log(f_{H_2}) - \frac{2.303RT}{F}\text{pH} \quad [1.2]$$

in which $\text{pH} = -\log(a_{H^+})$. The fugacity of hydrogen is related to the molal (mol/kg) concentration of dissolved hydrogen by Henry's Law as:

$$f_{H_2} = K_{H_2} \cdot \gamma_{H_2} \cdot m_{H_2} \quad [1.3]$$

where K_{H_2} and γ_{H_2} are Henry's constant and activity coefficient for dissolved hydrogen, respectively.

For the hydrogen electrode to be used as a reference electrode, the pH of the medium and the fugacity of the hydrogen must be accurately known. The first condition is readily achieved by employing a well-characterized buffer solution. Suitable buffers are available for temperatures up to ~300°C, primarily through the work of the Oak Ridge National Laboratory group, but few are available for use at temperatures up to the critical temperature. No buffers are available for use at supercritical temperatures; the few studies that have been performed in supercritical systems have employed HCl and NaOH solutions as pH standards, the pH values of which are readily calculated from dissociation constant data in the literature that were obtained from conductance studies. Note that both electrolytes are poorly dissociated at supercritical temperatures and hence are classified as being a weak acid and a weak base, respectively [8]. Furthermore, the corresponding fully dissociated salts, which would be required for an effective buffer to be formulated, do not exist. In most applications of the hydrogen electrode being used as a reference electrode, the major experimental problem is the accurate measurement of the hydrogen fugacity, as noted above. For instance, a temperature fluctuation of ±1°C at 300°C can easily cause an error in the estimated hydrogen partial pressure of greater than ±1 atm, which will be reflected in an uncertainty in the reference electrode potential of nearly 60 mV (if only the total pressure is known) [7]. Although values for the hydrogen fugacity coefficients are well known, it is generally necessary to employ extra-thermodynamic assumptions to estimate the fugacity coefficients in the presence of a high solvent (water) vapor pressure. For example, it is often assumed that partial pressures of the solvent (water) and hydrogen are additive, an assumption that has not been rigorously tested, to our knowledge.

The problem of the uncertainty in the hydrogen fugacity was overcome by the development in the early 1970s of the hydrogen concentration cell by Mesmer and Baes [9–13] for high temperature potentiometric studies. Various versions of the high temperature aqueous hydrogen concentration cell were subsequently developed by Mesmer *et al.* [9–14], Palmer and/or Wesolowski *et al.* [15–21] and Macdonald *et al.* [22–24]. Results using similar cells have also been reported by Shoesmith and Woon [25], Giasson and Tewari [26], Matsushima *et al.* [27], and Bilal and Mueller [28].

One classical example of the hydrogen concentration cell is shown in Fig. 1.7 [22]. The cell is formally represented as:

$$\text{Pt, } H_2/H^+(a_1)\text{, KCl} \mid H^+(a_2)\text{, KCl}/H_2\text{, Pt} \qquad [1.4]$$
(with liquid junction)

in which the vertical bar represents the liquid junction, the forward slash represents a phase boundary, and the comma separates two components in the same phase. The central compartment contains a buffered solution of accurately known pH–temperature characteristics and serves as the reference electrode, while the outer compartment contains the solution under study. The inner and outer compartments are connected via a porous Teflon plug, so that the cell contains a liquid junction, as indicated above, with the liquid junction potential being suppressed by using a high background concentration of KCl. Both compartments contain platinized platinum hydrogen electrodes and, during operation, the system is pressurized with

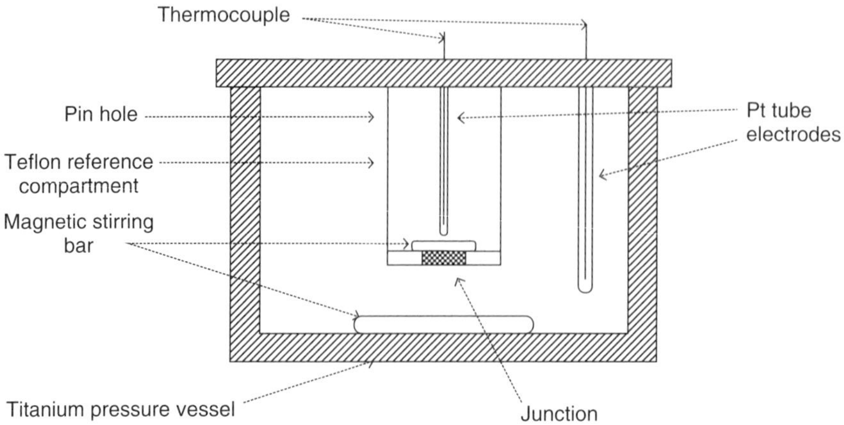

Fig. 1.7 Schematic of a high temperature aqueous hydrogen concentration cell [22].

hydrogen gas. Since the inner (reference) and outer (test) compartments are connected via a small hole above the liquid surface, the fugacity of hydrogen in both compartments is the same, but does not need to be known accurately because the terms containing the hydrogen fugacity in the Nernst equation cancel in the expression for the cell potential. Accordingly, the cell potential varies only with the change of the pH of the test solution.

One important variation of the hydrogen concentration cell, the 'flowing emf cell', was developed by Sweeton *et al.* [12] for studies of volatile reactants. The cell, shown in Fig. 1.8, comprises two electrode chambers and a liquid junction chamber. The lateral holes in the bottom of each piece allow solution to come in from each electrode chamber and then flow upwards into grooves cut in the connecting inside surfaces. The converging solutions are separated mechanically by a Teflon separator inserted between the two half-cylinders (the cylindrical liquid–junction plug). The two flowing streams join before leaving this split plug, then pass along a small groove cut along the outside surface of the upper plug to the outlet. The porous Teflon separator is intended to form and maintain a stable interface and to reduce

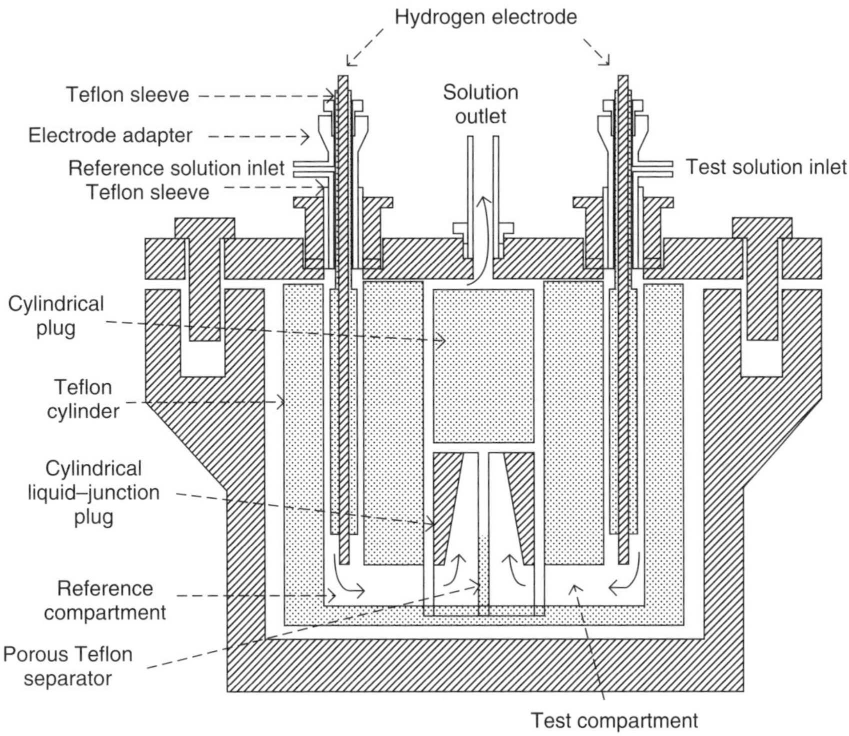

Fig. 1.8 Schematic cross-section of flowing hydrogen concentration cell [12].

convection that might result from slight differences in density of the two solutions and from heat released at the interface by the mixing of the solutions. This emf cell with a flowing liquid junction is capable of operating at pressures up to 100 atm and at temperatures up to 300°C.

One application of the cell is the determination of the dissociation quotient for water (Q_W). The cell used for this purpose can be represented as

$$\text{Pt, } H_2(m_1)/HCl(m_2),\ KCl(m_3) \mid KCl(m_3),\ KOH(m_2)/H_2(m_1),\ \text{Pt.} \quad [1.5]$$

The accuracy of the measured results was ±0.02 in log[Q_W] at temperatures ranging from 0 to 50°C. When the cell was tested as an HCl concentration cell (using different concentrations, c_1 and c_2, of HCl, instead of HCl and KOH, on both sides, see Eq. 1.4), with the potential between the electrodes being given as:

$$\Delta E = -\frac{2.303RT}{F}\log(c_2/c_1) + E_{LJ}, \quad [1.6]$$

the measured potential could be used to calculate the concentration ratio of HCl on the two sides. The measured ratios were found to be in good agreement with the true values over the full temperature range of 0 to 300°C [12].

The hydrogen concentration cell has been used to investigate the dissociation of water in KCl solution, to characterize buffer systems, including boric acid (50–290°C) [10], phosphoric acid (up to 300°C) [11], and ammonia (50–295°C) [13], to investigate the hydrolysis of Al^{3+} (up to 200°C) [9], to explore the precipitation of boehmite (γ-AlOOH) (200°C) [23], and to measure transport numbers for H^+ and Cl^- in aqueous HCl solutions at temperatures as high as 200°C [24], among other systems, where the upper temperatures are given in parentheses. One instrumental restriction on the applicable temperature of the cell comes from the use of PTFE, which is unstable at temperatures above 300°C. Hydrogen concentration cells have proven to be capable of precise in-situ pH measurement (with uncertainties of no greater than ±0.01 log unit), even over long periods of time (up to 1 month or more) [17]. Indeed, it is fair to say that essentially all of the high-precision pH measurements at moderate subcritical temperatures (100–300°C) reported to date have been performed using hydrogen concentration cells of the type first introduced by Mesmer and Baes [11].

Although the hydrogen concentration cell can provide precise pH measurement at temperatures up to 300°C, and possibly up to T_c, when employing a flowing cell [12], it has several inherent disadvantages:

1 It cannot be used in systems that are unstable in the presence of hydrogen.

2 It requires the use of a buffer solution whose pH–temperature characteristics are accurately known, which is difficult to fulfill under high subcritical temperatures and at supercritical temperatures, as noted above.
3 A suitably high background concentration of an indifferent electrolyte, such as KCl, must be employed to suppress the ILJP between the reference and test compartments.

Of these three issues, the first two are the most serious, with the first severely limiting the systems that can be studied to those that are stable in the presence of hydrogen, and the second limiting the upper temperature. The third constraint is not a major issue in high subcritical systems, because the transference numbers of the ions of most, if not all, binary electrolytes tend toward 0.5 with increasing temperature; however, at temperatures above the critical temperature the solubility of a salt is severely restricted and it may not be possible to attain a sufficiently high concentration to suppress the liquid junction potential. Note that the isothermal liquid junction is most effectively suppressed if the transference numbers of the cation and the anion of the background electrolyte are equal, a condition that is fulfilled by KCl at ambient temperature (and hence the reason for the choice of KCl in ambient temperature studies).

The hydrogen (reference) electrode has also been used extensively to explore the thermodynamics of the YSZME in high subcritical and in supercritical aqueous systems. The use of the YSZME in subcritical systems was first described in 1980 by Niedrach [29] and soon thereafter by Tsuruta and Macdonald [30], Danielson *et al.* [31, 32] and Bourcier *et al.* [33]. The YSZME with a Cu/CuO internal element was initially used as a reference electrode by Niedrach [29, 34–37] for measuring corrosion potentials of stainless steel components in nuclear reactor coolant circuits. The work of Macdonald *et al.* [6, 30, 38–41] and later by Lvov *et al.* [42] concentrated on exploring the thermodynamics of the YSZME, as noted above, because of its utility as a primary pH sensor (i.e., one that does not need to be calibrated). Since these initial studies, the YSZME with an Ag/O_2 internal element has been used to sense acidity in aqueous solutions at temperatures as high as 528°C using an external pressure-balanced reference electrode (EPBRE) based upon the Ag/AgCl, KCl (sat), electroactive element [43, 44]. This work is described further below. Regardless of whether the YSZME is employed as a reference electrode (in a medium of known pH versus temperature characteristics, as in the initial application by Niedrach [29, 34–37]) or as a pH sensor, it is necessary to establish the theoretical response of the electrode in order that the potential can be related to the SHE scale.

For the YSZME electrode employing an Hg/HgO internal element, the half-cell reaction is written as:

$$HgO + 2H^+ + 2e^- \Leftrightarrow Hg + H_2O \qquad [1.7]$$

with the corresponding Nernst equation being:

$$E_{YSZ} = E^0_{Hg/HgO} - \frac{2.303RT}{2F}\log(a_{H_2O}) - \frac{2.303RT}{F}\mathrm{pH} \qquad [1.8]$$

where $E^0_{Hg/HgO}$ is the standard potential of the Hg/HgO couple and a_{H_2O} is the activity of water. By combining Eq. 1.2, 1.3 and 1.8, we obtain the potential of the cell H_2/H^+, YSZME as:

$$\Delta E = \left[E^0_{Hg/HgO} + \frac{2.303RT}{2F}\log\left(\frac{K_H \cdot \gamma_{H_2}}{a_{H_2O}}\right)\right] + \frac{2.303RT}{2F}\log(m_{H_2}). \qquad [1.9]$$

The first term of Eq. 1.9 is a constant for a given temperature, pressure and ionic strength, so that the cell voltage is linearly related to the logarithm of the dissolved hydrogen concentration. In other words, using the hydrogen electrode as an indicator electrode and the YSZME as the reference electrode, the cell

$$\mathrm{Pt},\ H_2/H^+(aq.)/ZrO_2\ (Y_2O_3)/HgO/Hg \qquad [1.10]$$

may be employed as a hydrogen sensor to obtain the concentration of hydrogen at given temperature and pressure by measuring the potential difference between the two electrodes [6, 40]. To test this concept, Ding and Seyfried Jr [45] used the same electrochemical cell illustrated above to measure the fugacity and Henry's constant for hydrogen at supercritical temperatures. [Dr Ding was instructed on the design, construction and operation of the cell during a multi-day visit to the author's (D.D.M.'s) laboratory. At that time, the cell was being used by graduate student, K. Eklund, to measure Henry's constant for hydrogen in NaOH solutions at supercritical temperatures.] At a temperature of 400°C and at a pressure of 400 bar, the measured potential of the cell displays a good Nernstian response to dissolved hydrogen concentration that agrees with the theoretical value. The experiments also demonstrated the excellent stability of the sensor during month-long operation [45].

Eklund *et al.* [46] also employed the hydrogen electrode and YSZ electrode to measure Henry's constant for hydrogen in NaOH solutions at temperatures ranging from 25 to 450°C and at a pressure of 275 bar. The system that they used can be written in the form:

$$Hg/HgO/ZrO_2\ (Y_2O_3)/NaOH(aq.)/H_2,\ \mathrm{Pt}. \qquad [1.11]$$

The results agreed well with those of Fernandez-Prini and Crovetto [47] for temperatures up to 350°C and with Kishima and Sakai's work [48] for temperatures up to 450°C.

According to Ding and Seyfried Jr [49], two problems of the platinum/hydrogen electrode are the poisoning by sulfide ion and the solubility of hydrogen in the metal at elevated temperatures. We regard thc first of these concerns to be legitimate, but it is difficult to reconcile the second with thermodynamic principles. Ding and Seyfried Jr [49] evaluated gold hydrogen-sensing electrodes, as alternatives to platinum, for the in-situ measurement of dissolved hydrogen in supercritical aqueous fluid (temperature: 375–400°C, pressure: 400 bar). With the same cell potential expressed by Eq. 1.9, at 400°C and 400 bar, the experimental data reveal a slope of 0.0668 ± 0.0027 V (measured potential with respect to logarithm of dissolved H_2 concentration), which is in excellent agreement with the theoretical values of 0.0668 V (from Eq. 1.9) compared with that for a platinum electrode under the same conditions (0.054 V, see Ref. 45). The present authors consider the latter result to be too far different from the theoretical value that attribution of the difference to the different noble metals employed is not reasonable. Instead, the difference is probably due to experimental error, given the high level of difficulty in performing experiments of this type.

A significant problem with potentiometric measurements in high subcritical and in supercritical aqueous systems is corrosion of the apparatus, which can result in large changes in the pH. This problem was addressed by Lvov *et al.* [50–52] by devising flow through reference electrodes (Ag/AgCl and the YSZME) and indicator electrodes (H_2/H^+), such that the incoming solution was never in contact with corrodible metal surfaces. In this way, contamination of the solution could be prevented or at least minimized. Sue *et al.* [53] subsequently employed this concept by using two identical high temperature platinum/hydrogen electrodes as the working and reference electrodes (Fig. 1.9). The cell is represented as:

$$\text{Cu/Pt, H}_2\text{/Ref. Solution} \mid \text{Test Solution/H}_2\text{, Pt/Cu} \qquad [1.12]$$

$$T_1 \rightarrow \mid \leftarrow \qquad T_2 \qquad \rightarrow \mid \leftarrow T_1$$

in which T_2 is the higher working temperature and T_1 is the ambient reference temperature. The corresponding cell potential is written as

$$E = \frac{1}{F}[\mu_{H^+}(L) - \frac{1}{2}\mu_{H_2}(L) + \frac{1}{2}\mu_{H_2}(R) - \mu_{H^+}(R)] + \Delta\Phi_{LJ} + \Delta\Phi_{STR}(L) + \Delta\Phi_{STR}(R) \qquad [1.13]$$

in which μ_{H^+} and μ_{H_2} are the chemical potentials of H^+ and H_2, respectively, $\Delta\Phi_{LJ}$ and $\Delta\Phi_{STR}$ are the isothermal liquid junction potential and the steaming potentials, and L and R designate the left and right side of the cell, respectively. In a manner similar to that previously described by Lvov *et al.*

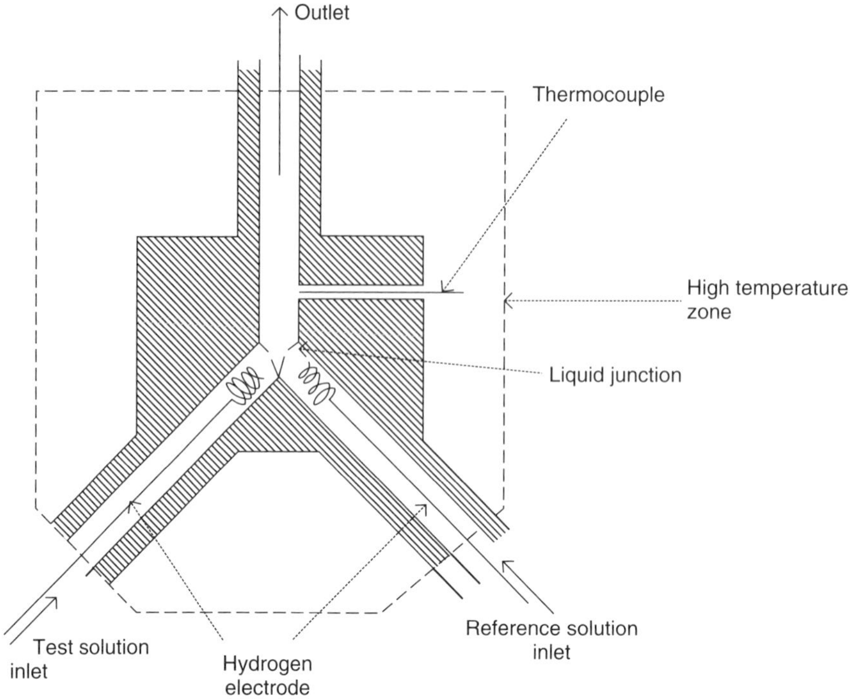

Fig. 1.9 Schematic of flow-through electrochemical cell [53].

[50–52], measurement of the cell potential for Test Solutions 1 and 2 yields the expression:

$$\Delta E_{ap} = \frac{RT}{F} \ln \frac{m_{H^+,test(2)}\gamma_{\pm,test(2)}}{m_{H^+,test(1)}\gamma_{\pm,test(1)}} + \Delta\Phi_{LJ} \quad [1.14]$$

in which m_{H^+} and $\gamma_\pm$ are the hydrogen ion molality and the mean ionic activity coefficient, respectively. The mean molal activity coefficient, $\gamma_\pm$, can be calculated using extended Debye–Huckel theory and $\Delta\Phi_{LJ}$ can be estimated by using the Henderson equation. They also described procedures to reduce the instability of the measured potential, including adjustment of the electrode coiled tip lengths, etc., although the exact theoretical basis for the procedures remains obscure. Sue *et al.* [53] compared the calculated and measured values for ΔE_{ap} and concluded that an accuracy of ±0.02 of a logarithmic unit was achieved at temperatures from 19.6 to 392.9°C and at pressures from 0.1 to 29.8 MPa. The cell was also used to determine the first dissociation constant of sulfuric acid at temperatures from 375.2 to 399.8°C and at pressures from 28.1 to 32.5 Mpa [54].

1.2.2 Silver/silver chloride electrodes

Silver/silver chloride electrodes have been used extensively as reference electrodes in high temperature electrochemical studies. The first potentiometric measurement of pH at elevated temperatures, at Oak Ridge National Laboratory by Lietzke [55], employed Pt-H_2 electrodes coupled with Ag/Ag-salt reference electrodes to investigate the thermodynamic properties of HCl and other strong acids in electrolytes to 225°C. Ag/AgCl electrodes are considered to be among the most serviceable electrodes under high temperature conditions, and have been found to perform satisfactorily at temperatures up to 573 K (300°C) [7]. Thus, the early work of Greeley and Lietzke [56–58], subsequently reviewed by Macdonald and Kriksunov [7], attests to the utility of the Ag/AgCl electroactive element, in particular.

One of the major problems experienced with all Ag/AgX, X = Cl, Br, I electrodes, when used in hydrogen-rich systems, is the reduction of AgX by hydrogen to metallic silver:

$$AgX + \tfrac{1}{2}H_2 \rightarrow Ag + H^+ + X^-. \qquad [1.15]$$

This reaction results in a mixed potential (rather than an equilibrium potential) being generated at the electroactive element and in the activity of X^- at the electroactive element varying with time. Further, as noted above, the thermohydrolysis of AgX:

$$2AgX + H_2O \rightarrow Ag_2O + 2H^+ + 2X^- \qquad [1.16]$$

readily occurs at elevated temperatures and the reaction product will also result in a mixed potential. Furthermore, Ag_2O decomposes to Ag at temperatures greater than 160°C [59], so that the ultimate product of thermohydrolysis is silver metal. Both phenomena preclude true thermodynamic equilibrium, thus the results obtained from Ag/AgCl electrodes under those conditions should be regarded with caution.

Ding and Seyfried Jr [60] used the Ag/AgCl internal reference electrode in the direct measurement of pH in NaCl-bearing fluids at 400°C and 40 MPa. The electrochemical cell that was employed can be described as:

$$Ag/AgCl/Cl^-(aq.),H^+(aq.)/ZrO_2(Y_2O_3)/HgO/Hg \qquad [1.17]$$

with the cell potential being related to the pH of the solution by the equation:

$$\Delta E(V)_{T,p} = \Delta E^0 - \frac{2.303RT}{F}[\log a(Cl^-) - \frac{1}{2}\log a(H_2O)] + \frac{2.303RT}{F}pH_{T,p}. \qquad [1.18]$$

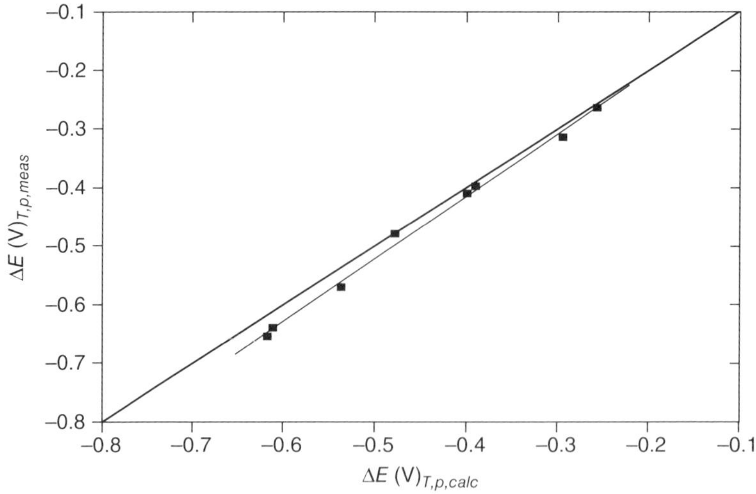

Fig. 1.10 Measured cell potential $\Delta E(V)_{T,p,meas}$ versus theoretically predicted cell potential, $\Delta E(V)_{T,p,calc}$, in the aqueous NaCl–HCl system (NaCl 0.57 M) [thermocell Ag/AgCl/Cl^-(aq.),H^+(aq.)/$ZrO_2(Y_2O_3)$/HgO/Hg, Eq. 1.17] at 400°C and 40 MPa [60].

Ding and Seyfried Jr [60] measured the pH of concentrated NaCl–HCl fluids (0.57 M NaCl) and compared the results with calculated cell potential data using various thermodynamic databases and codes [61–66]. The comparisons are shown in Fig. 1.10 and 1.11. The good agreement observed between $\Delta E(V)_{T,p,meas}$ and $\Delta E(V)_{T,p,calc}$ shown in Fig. 1.10 demonstrates that the cell response is consistent with available thermodynamic data. Furthermore, Fig. 1.11 shows an excellent linear relationship between $\Delta E(V)_{T,p,meas}$ and $pH_{T,p,calc}$ over the entire range investigated. Importantly, the slope of the regression line is 0.140 ± 0.004 V, which is quite close to the value of 0.134 V predicted from the Nernst equation for a temperature of 400°C.

1.2.3 Other internal reference electrodes

Other internal reference electrodes have been devised for use in high subcritical and in supercritical aqueous solutions. For example, Dobson *et al.* [67–69], Macdonald *et al.* [70, 71], and Nagy and Yonco [72] explored the palladium/hydrogen system as a reference electrode as well as a pH indicator electrode. The work by Dobson [67–69] and Macdonald [71] used palladium wire electrodes that had been cathodically charged into the β region of the Pd-H phase diagram, where solid palladium hydrides are formed. Hydrogen was then allowed to desorb from the lattice and the system passed into the $\alpha + \beta$ region, where hydrogen in solid solution

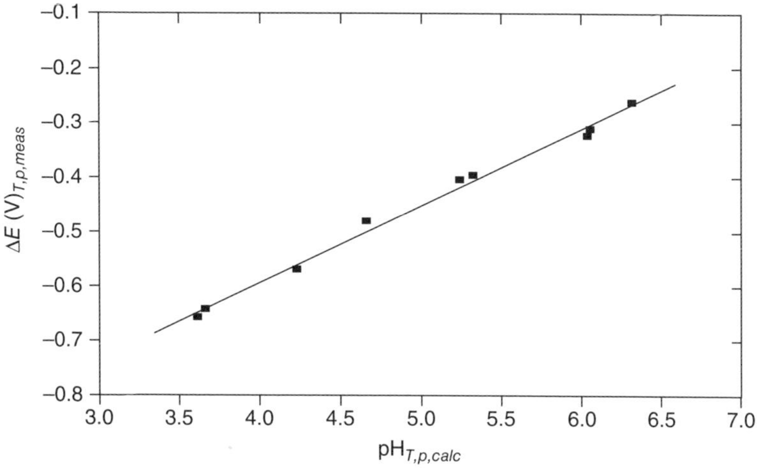

Fig. 1.11 Measured cell potential $\Delta E(V)_{T,p,meas}$ against theoretically predicted $pH_{T,p,calc}$ [thermocell Ag/AgCl/Cl^-(aq.),H^+(aq.)/$ZrO_2(Y_2O_3)$/HgO/Hg, Eq. 1.17] at 400°C and 40 MPa [60]. The buffer solution is NaCl–HCl aqueous system (NaCl 0.57 m), pH is varied by adding HCl.

(the α phase) is in equilibrium with the hydrides. Under these conditions, the activity of H in the lattice is fixed, so that the potential due to the reaction $H^+ + e^- \Leftrightarrow H$ depends only on the pH and temperature, provided that the composition remains in the $\alpha + \beta$ region of the phase diagram. Once the composition moves into the α region, where the activity of H depends on composition, the potential varies with time. Thus, as the potential decays from the cathodically charged state, it exhibits a plateau over which the activity of H in the lattice is established internally. This removes one of the principal objections to the hydrogen electrode as a reference electrode: the need to accurately control the hydrogen fugacity. However, the Pd-H phase diagram exhibits an upper consolute temperature for the $\alpha + \beta$ region of the phase diagram of about 275°C [71]. At temperatures above this value, the $\alpha + \beta$ region no longer exists, so that the establishment of a fixed activity of H does not occur.

A variant of the Pd-H_2 electrode was explored by Macdonald *et al.* [70] and later by Nagy and Yonco [72] in the form of a Pd (or Pd-Ag) thimble with a known H_2 pressure on the inside with the external surface being exposed to the solution. Diffusion of hydrogen through the Pd wall establishes a fixed activity of H (or fugacity of H_2) at the outer surface where the potential determining reaction occurs. The electrode potential is stable and the electrode can serve as a viable reference electrode, provided that the pH is known. Again, like the Pd-H_2 electrode, the Pd thimble electrode removes the difficulty with the classical hydrogen electrode of establishing a known hydrogen activity (fugacity) in the system. However, unlike the

Pd-H_2 electrode, the maximum temperature of operation is not limited by the thermodynamics of the Pd-H system. Of course, these reference electrodes are viable only in systems that are stable in the presence of hydrogen.

Finally, attempts have been reported to develop various solid state electrolyte internal reference electrodes for use in high subcritical and supercritical aqueous systems. Among these was an attempt by Hettiarachchi and Macdonald [73] to develop a solid polymer electrolyte (epoxy + KCl solution, which sets as a rubber) reference electrode containing an Ag/AgCl electroactive element (Fig. 1.12 [73]). The electrode displayed reversible behavior at temperatures as high as 280°C for more than 40 h, but, because the activities of Cl^- in water-containing epoxy are not known, the electrode needs to be calibrated. A variation on this theme was reported by Jayaweera *et al.* [74] who used a zirconia + alumina + KCl + silica electrolyte, together with an Ag/AgCl electroactive element, as a reference electrode for supercritical systems. The electrode was found to exhibit a constant potential, but it suffers from the same disadvantage as does the polymer electrolyte

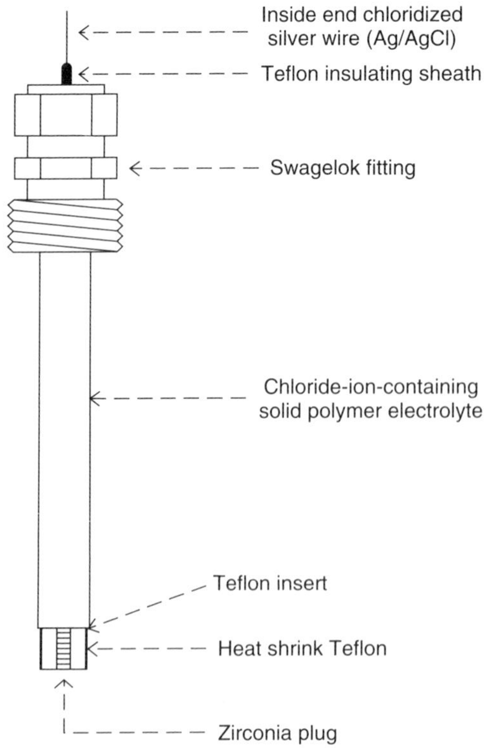

Fig. 1.12 Assembly of an internal polymer electrolyte reference electrode [73].

reference electrode: the need to calibrate, which requires a primary standard whose potential can be accurately related to the SHE scale.

1.3 External reference electrodes

Because of the harsh working conditions in high subcritical and in supercritical water systems, the stability of electroactive element becomes a major practical and theoretical issue in reference electrode technology. The most frequent problem is the thermal hydrolysis of one or more phases in the electroactive element, such as AgCl in the Ag/AgCl internal reference electrode, as described above. One solution to this problem is to locate the electroactive element at ambient temperature, where thermal hydrolysis does not occur, and connect the element to the high temperature region of the system by a nonisothermal electrolyte bridge. Early variants of this scheme used a nonisothermal bridge together with a compressed porous plug to reduce the pressure to ambient at the electroactive element (de Jones and Masterson [75], Macdonald and Owen [76], and Macdonald [70]). Accordingly, two irreversible potentials are created by the junction. The first is the steaming potential due to the nonisobaric junction, which is eliminated by the use of pressure-balanced electrolyte bridges. The second is the thermal diffusion potential, which remains as the main problem of all external pressure-balanced electrodes.

External pressure-balanced reference electrodes (EPBREs) have been developed and evaluated extensively over the past three decades. The first variant was described in 1978 by Macdonald *et al.* [77] and was patented in the same year, with the invention being assigned to the Electric Power Research Institute. The electrode won the EPRI 'Invention-of-the-Year Award' in 1979. The EPBRE that was first evaluated is described thermodynamically as the thermocell:

$$(25^{\circ}C)\ Ag/AgCl/KCl\ (M)/AgCl/Ag\ (T^{\circ}C) \qquad [1.19]$$

in which T was varied from 25 to 275°C and the concentration of KCl was varied from 0.0050 to 0.505 M [77]. An important feature of the original, Macdonald *et al.* [77] electrode was the reliance of pressure pulsations transmitted into the flexible (PTFE) inner compartment from the flow-activating pump, in order to inhibit the tendency of thermal diffusion to occur and hence to maintain thermal diffusion in the Soret initial state (Fig. 1.13). This was an important, but often ignored, caveat of the operation of an EPBRE. If thermal diffusion occurs, the potential will drift as the electrolyte diffuses toward the cold end of the junction (the tendency predicted by thermal diffusion theory). The theory of thermal diffusion of the electrolyte applies strictly to subcritical systems, where the properties of the system can be described in terms of classical electrolyte theory. No

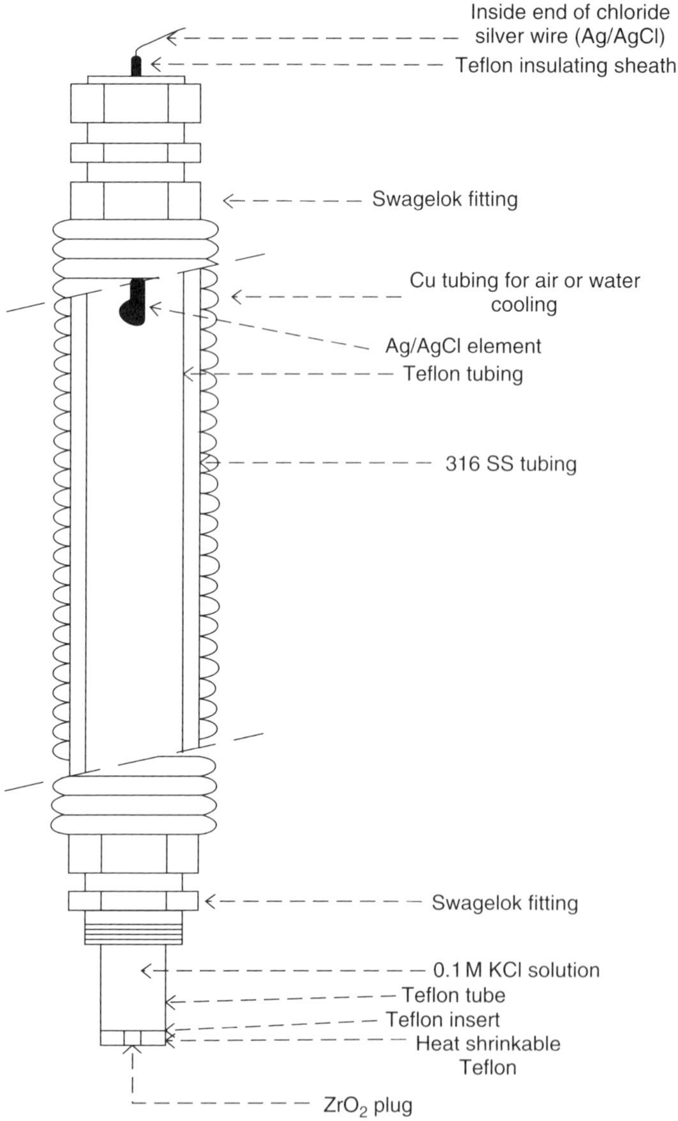

Fig. 1.13 Assembly of an external pressure-balanced reference electrode [39].

comparable theory has been developed for supercritical aqueous systems, to our knowledge. In spite of the theoretical shortcomings, EPBREs have been used extensively in supercritical aqueous systems to demonstrate the thermodynamic viability of YSZMEs [6, 40, 78] and to measure pH in high subcritical and in supercritical aqueous solutions (using YSZMEs as pH electrodes) [39, 73, 79, 80]. In the early 1990s, using an YSZME with an Ag/O_2 internal element and an EPBRE, Kriksunov and Macdonald significantly extended the temperature of direct acidity measurement up to 528°C [41, 43, 44]. To our knowledge, this still represents the highest temperature at which acidity measurements have been made.

The original flow-through technique for reference electrodes was developed by Danielson in 1983 [81] with the purpose of eliminating thermal diffusion by continually refreshing the inner compartment. In this way, the electrolyte in the inner compartment could be maintained in the Soret initial state (i.e., by maintaining a uniform concentration along the nonisothermal bridge). The flow rate was adjusted to be sufficiently high to maintain a uniform concentration, but to be sufficiently low to maintain the temperature gradient along the nonisothermal electrolyte bridge. The original Danielson cell, which was restricted to moderate, subcritical temperatures ($T < 300$°C), proved to be quite successful, in that the potential remained stable at the expected value over extended periods of time. Later, this same concept was adopted by Lvov *et al.* (Fig. 1.14) [50, 51], for high subcritical and low supercritical temperatures (374.15°C < T < 400°C), with the result that highly accurate potentiometric measurements could be made [50, 51, 82]. This required careful calibration of the reference electrode against a hydrogen electrode in a flow-through system and against theoretical calculation, knowing the pH and the fugacity of hydrogen. In any event, while this type of electrode represents the current 'state-of-the-art', it is judged to be too difficult to apply in the field and is perhaps too difficult to apply even in the research laboratory (the same problem with the EPBRE). The reference potential was also found to depend on flow rate of the electrolyte through the inner compartment, presumably arising from the streaming potential and possibly also from the modified temperature gradient down the thermal liquid junction. The fact that the potential is flow rate dependent is a significantly complicating factor, because of the need to calibrate the electrode in each system studied.

As noted above, an external pressure-balanced electrode suffers from two irreversible thermodynamic potentials: the isothermal liquid junction potential and the thermal liquid junction potential (thermal diffusion potential). The EPBRE is amenable to formal thermodynamic analysis in terms of Thermocell I [51]:

$$\begin{array}{l} |\leftarrow \ T_1 \ \rightarrow|\leftarrow \quad T_2 \quad \rightarrow|\leftarrow \ T_1 \ \rightarrow| \\ \mathrm{Cu/Ag/AgCl/KCl\ (aq.)\ |\ HCl\ (aq.)/H_2,\ Pt/Cu.} \end{array} \qquad [1.20]$$

The initial emf (corresponding to the Soret initial state) of the cell is given by (i.e., no concentration gradient)

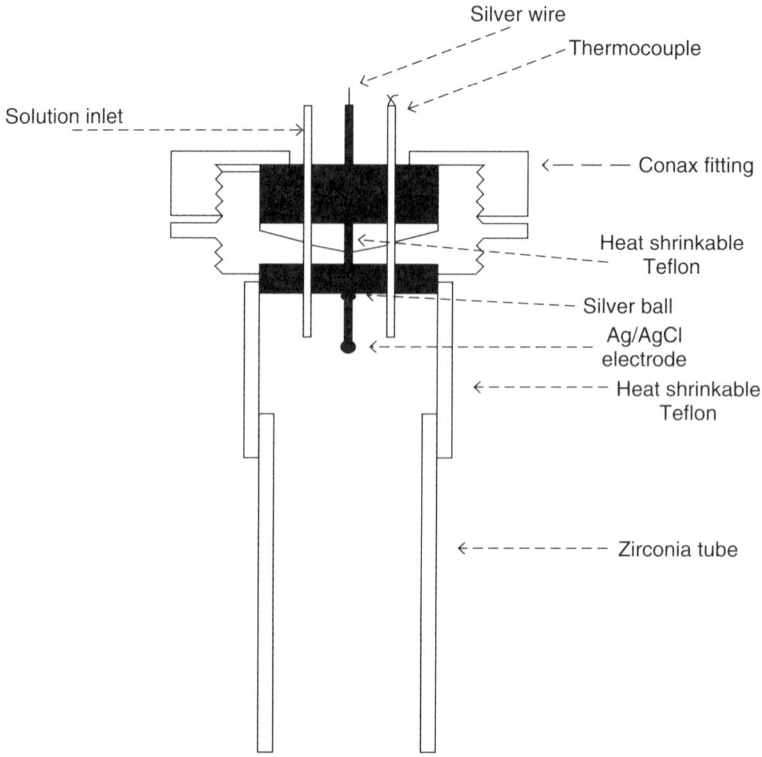

Fig. 1.14 Schematic of the flow-through external pressure-balanced reference electrode [51].

$$E = -\frac{1}{F}\int_{T_1}^{T_2} \bar{\bar{S}}_{\mathrm{e,Pt}}\,\mathrm{d}T + E_{\mathrm{TLJ}} + E_{\mathrm{D}}(T_2) + \frac{1}{F}[\mu_{\mathrm{H^+}}(T_2) + \mu_{\mathrm{Cl^-}}(T_1) - \frac{1}{2}\mu_{\mathrm{H_2}}(T_2) - \mu_{\mathrm{AgCl}}(T_1) + \mu_{\mathrm{Ag}}(T_1)] \qquad [1.21]$$

in which $\mu_{H^+}(T_2)$ and $\mu_{H_2}(T_2)$ are the chemical potentials of the hydrogen ion and the hydrogen molecule in the high temperature (T_2) region, respectively. $\mu_{Ag}(T_1)$, $\mu_{AgCl}(T_1)$, $\mu_{Cl^-}(T_1)$ are the chemical potential of Ag, AgCl and Cl^- in the low temperature (T_1) zone, respectively. The term $\bar{\bar{S}}_{e,Pt}$ is the transported entropy of an electron in the platinum wire. $E_D(T_2)$ and E_{TLJ} are, respectively, the isothermal liquid junction potential and thermal liquid junction potential, which are discussed below.

1.4 Isothermal liquid junction potential

Both internal and external reference electrodes possess an interface between the internal solution and the external environment. This interface is commonly established within a porous junction and is designed to permit electrolytic communication while preventing flow. In any event, the junction gives rise to the isothermal liquid junction potential (ILJP), $E_D(T_2)$, which develops because some ions diffuse faster than others, thereby generating an electric field that opposes the process. Integration of the electric field across the junction yields the ILJP. Bard and Faulkner [83] provide a detailed discussion of the thermodynamics of the junction [83]. For dilute solutions, the potential can be calculated from Henderson's equation. In the case of Thermocell I, the ILJP is expressed by:

$$E_D(T_2) = \frac{\sum_i \frac{|z_i|u_i}{z_i}(m_i^{HCl} - m_i^{KCl})}{\sum_i |z_i|u_i(m_i^{HCl} - m_i^{KCl})} \frac{RT_2}{F} \ln \frac{\sum_i |z_i|u_i m_i^{KCl}}{\sum_i |z_i|u_i m_i^{HCl}} \qquad [1.22]$$

where u_i and z_i are the mobility of and charge on species i, respectively. The quantities m_i^{KCl} and m_i^{HCl} are the molal concentration of the ith species of the right (HCl) and left (KCl) sides of the junction. The ILJPs for various $B(OH)_3$/LiOH aqueous systems at a number of temperatures have been calculated by Macdonald *et al.* [71] and have been used to correct the measured results in Ref. 84.

The Henderson's equation assumes ideal solution behavior, which renders it suitable only for dilute cases. For more concentrated solutions, Harper has provided an alternative expression [85]. The equation is reckoned to achieve an accuracy of 0.1 mV for a simple junction between two identical electrolytes of different concentrations, and 1–2 mV uncertainty for more complex junctions.

As noted above, the ILJP can be suppressed by employing a high concentration of a binary electrolyte whose ions have similar transference numbers. Traditionally, in ambient temperature studies, KCl has been used for this purpose, because of the near equality of the transference numbers: $t^0_{K^+} = 0.491$ and $t^0_{Cl^-} = 0.509$ at 25°C. As shown by Macdonald and Owen [24], the transference numbers for H^+ and Cl^- approach equality as the temperature increases, and ionic conductance data indicate that this is a general trend, suggesting that the ILJP issue is of less importance at higher temperatures. However, the trend toward equality of the transference numbers is countered by the larger value of RT/F, so that the effect of temperature on the ILJP is not easily gleaned by inspecting Eq. 1.22 alone.

1.5 Thermal liquid junction potential

E_{TLJ} in Eq. 1.21 is the thermal liquid junction potential (TLJP), or the 'thermal diffusion potential'. In the electrolyte solution, a small, but definite, increase in concentration occurs at the cold end of a nonisothermal electrolyte bridge as the system evolves towards a steady state. This transport phenomenon is denoted as 'thermal diffusion' and is due to coupling between the heat flux from the hot end to the cold end and ion flux. Thermal diffusion in liquid systems was first discovered by Ludwig in 1856, but was examined more closely in 1879–81 by Soret, and so is now named the 'Soret effect' [86]. The state when no obvious thermal diffusion has occurred is designated the 'Soret initial state' [87]. Concentration gradients due to the Soret effect lead to the establishment of a diffusion potential and an internal electric field due to the migration of ions. The sum of these two potentials is designated the thermal diffusion potential. Thermal diffusion continues until backwards 'chemical diffusion' (from the cold end to the hot end of the junction) causes the system to attain a steady state. This state is referred to as the 'Soret final state'. Attainment of the Soret steady state may take many days or weeks, depending upon the electrolyte, temperature gradient, and the length of the junction. Clearly, it is impractical to operate an EPBRE in the Soret steady state, so that emphasis has been placed on devising methods for maintaining the system in the Soret initial state (e.g., by transmitting pressure pulses into the inner compartment of an EPBRE, as discussed above). The importance of thermal diffusion in electrochemical cells has been discussed at length by Engelhardt *et al.* [88].

In 1979, Macdonald *et al.* [89] successfully measured the initial TLJP of the thermal cell

$$(298.15\,\text{K})\text{Ag/AgCl/KCl(aq.)/AgCl/Ag}(T) \qquad [1.23]$$

in which the temperature of the hot end of the cell (T) was varied from 298.15 to 548.15 K and the KCl concentration was varied from 0.005 to 0.505 mol/kg. Macdonald *et al.* calculated the Nernstian thermal cell potential by employing Gibbs energies of formation for the cell components based upon an absolute, 298.15 K, standard state. The difference between the Nernstian potential and the observed potential (Fig. 1.15) is the initial state TLJP (Fig. 1.16). The TLJP was found to increase in a parabolic manner with the temperature difference between T and 298.15 K (in Ref. 89, E_{TLJ} is represented as a cubic equation in ΔT, the temperature difference between the hot and cold ends of the cell). Another important feature of the measured TLJP is that it is only weakly dependent on concentration. Other studies on measuring or calculating TLJPs are described in Ref. 70 and 90.

Lvov and Macdonald [87] attempted to calculate the TLJP of Thermal Cell I from fundamental theory using the general expression for the potential as given by:

$$E_{TLJ} = -\frac{1}{F}\int_{T_1}^{T_2}\sum_i \frac{t_i S_i^*}{z_i}\,\mathrm{d}T \qquad [1.24]$$

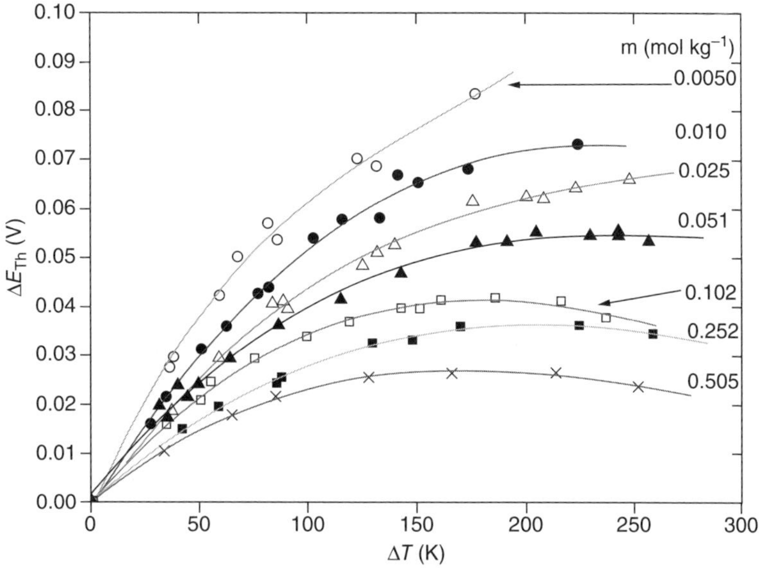

Fig. 1.15 Voltages for the thermocell (298.15 K)Ag/AgCl/KCl(aq.)/AgCl/Ag(T) (Eq. 1.23) as a function of $\Delta T[=(T-298.15)\text{K}]$ and KCl concentration (molal: m) [89]. The values shown on the right side of the figure are the KCl concentrations in mol kg^{-1}.

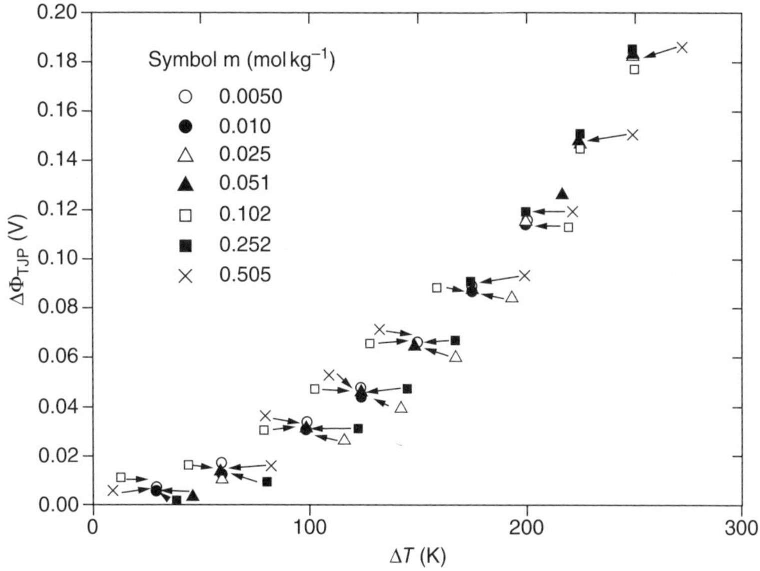

Fig. 1.16 Thermal liquid junction (TLJ) potentials for the thermocell (298.15 K)Ag/AgCl/KCl(aq.)/AgCl/Ag(T) (Eq. 1.23) as a function of ΔT $[=(T-298.15)\text{K}]$ for various KCl concentrations [89]. The values shown on the left side of the figure are the KCl concentrations in mol kg^{-1}.

in which t_i, S_i^* and z_i are the transport number, the entropy of transport, and the charge of the ith ion, respectively, in the thermal bridge solution. To calculate the thermal diffusion potential, the following information is required.

1 The ionic composition of the thermal liquid junction, in terms of the total composition of the solution at a given temperature and pressure.
2 The transference numbers, t_i, of all ions in the aqueous solution at a given temperature and pressure.
3 Eastman entropies of transport, S_i^*, of all ions in the solution.

First, Lvov and Macdonald used a free-energy minimization code [91], which employs standard Gibbs energies of formation of the species (as calculated using the SUPCR92 software package [92]). The activity coefficients for all charged species were calculated using extended Debye–Huckel theory, and activity coefficients for all neutral species, including water, were set equal to unity. Secondly, the ionic transference numbers are calculated using the equation:

$$t_i \approx \frac{|z_i|\lambda_i^0 c_i}{\sum_i |z_i|\lambda_i^0 c_i} \quad [1.25]$$

where λ_i^0 is the limiting equivalent ionic conductivity of the ith ion. Lvov and Macdonald used the ionic limiting equivalent conductivities of K^+(aq.), Na^+(aq.), Cl^-(aq.), H^+(aq.) and OH^-(aq.) given by Quist and Marshall [93]. The parameter c_i is the molar concentration of the ith species, which was calculated as described above. Equation 1.25 was proved to be a very good approximation for dilute solutions.

To derive the entropies of transport for ionic species, Lvov and Macdonald [87] extrapolated the standard entropies of transport of electrolytes, $S_i^{*,0}$, from lower temperatures to higher temperatures by applying Agar's hydrodynamic theory. The low temperature experimental data were taken from Ref. 94 and 95. Lvov and Macdonald [87] employed previously measured differences between E_{TLJ} values for different electrolytes, MCl_n(aq.), so as to test the viability of the extrapolation. The results are shown in Fig. 1.17. As can be seen, good agreement exists between the calculated and observed data. Lvov and Macdonald [87] also calculated TLJPs for different electrolytes over a wide range of temperatures (Fig. 1.18 and 1.19). From the calculated data, it can be seen that E_{TLJ} depends strongly on both the temperature and the type of the electrolyte, and the difference could be more than ±150 mV. To our knowledge, these data represent the only reported attempt to estimate the TLJP for EPBREs.

All of the calculations described above correspond to Soret's initial state, i.e., to a state where the concentration of electrolyte along the bridge is

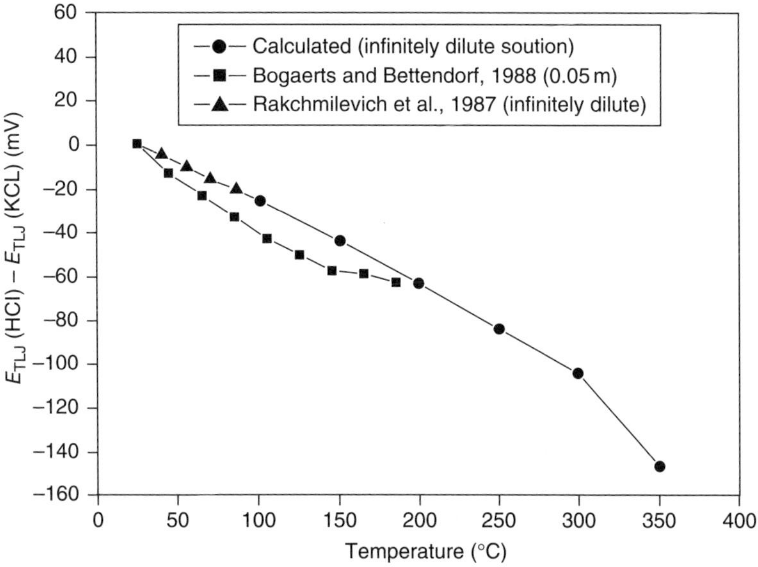

Fig. 1.17 Comparison of the calculated and observed E_{TLJ}(HCl) – E_{TLJ}(KCl) values (Thermocell I, Eq. 1.20) as a function of temperature [87].

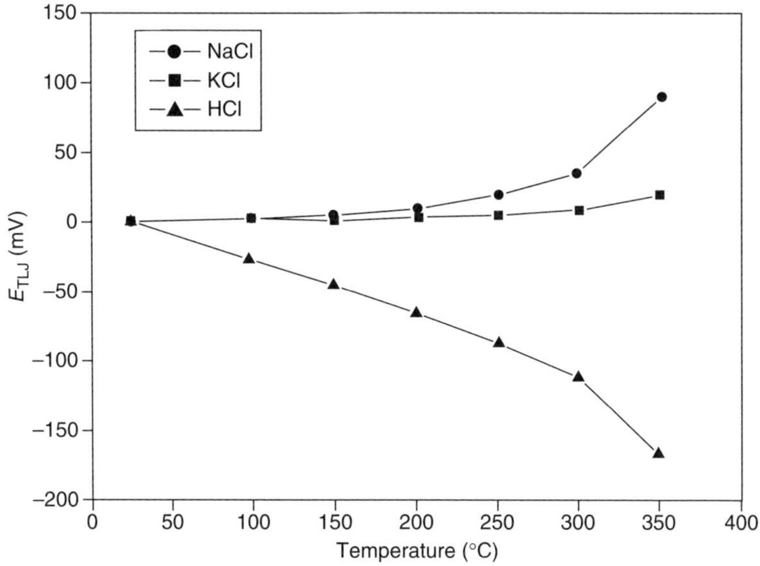

Fig. 1.18 Temperature dependence of E_{TLJ} (Thermocell I) for standard, infinitely dilute NaCl, KCl and HCl aqueous solutions [87].

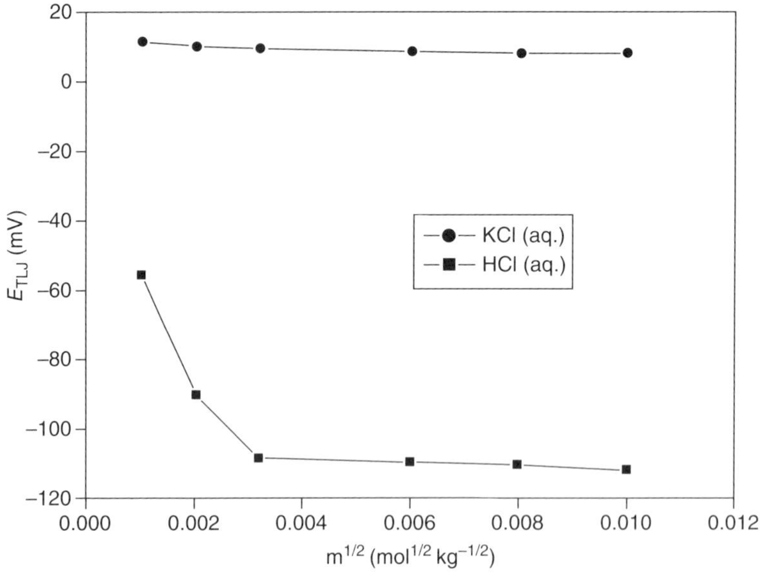

Fig. 1.19 Concentration dependence of E_{TLJ} (Thermocell I) for very dilute aqueous KCl and HCl solutions at 300°C [87].

constant. As noted above, maintaining the thermal junction in the initial state has been a vital problem in the operation of EPBREs. As also noted above, in the original design of the EPBRE [77] (Fig. 1.13), the thermal junction consisted of a PTFE tube and a porous ZrO_2 plug at the end. The PTFE tube was also used to transmit the pressure pulses from the cell to the internal reference solution. Thus any thermal diffusion tendency was counteracted by the pressure pulse. When the EPBRE was used under supercritical conditions [6, 39, 40], a ceramic bridge was used in the high temperature ($T > 300$°C) part of the junction, and it was connected via a PTFE tube to the lower temperature region (since the highest temperature PTFE can sustain is 563 K). The uncertainty of the electrode potential was measured to be ±15–25 mV at temperatures below 573 K, and up to 100 mV at supercritical temperatures, and the measured potential was found to be extremely sensitive to pressure changes.

A variation of the external reference electrode concept was reported by Tsuruta and Macdonald [84] to measure the pH of a high temperature solution (Fig. 1.20). The ceramic membrane pH-measuring system is represented by

$$|\leftarrow T_1(25°C) \rightarrow|\leftarrow \quad T \quad \rightarrow|\leftarrow T_1(25°C) \rightarrow|$$

$$\text{Ag/AgCl/Internal Solution/Test Solution/}ZrO_2(Y_2O_3)\text{/Internal Solution/Ag/AgCl} \qquad [1.26]$$

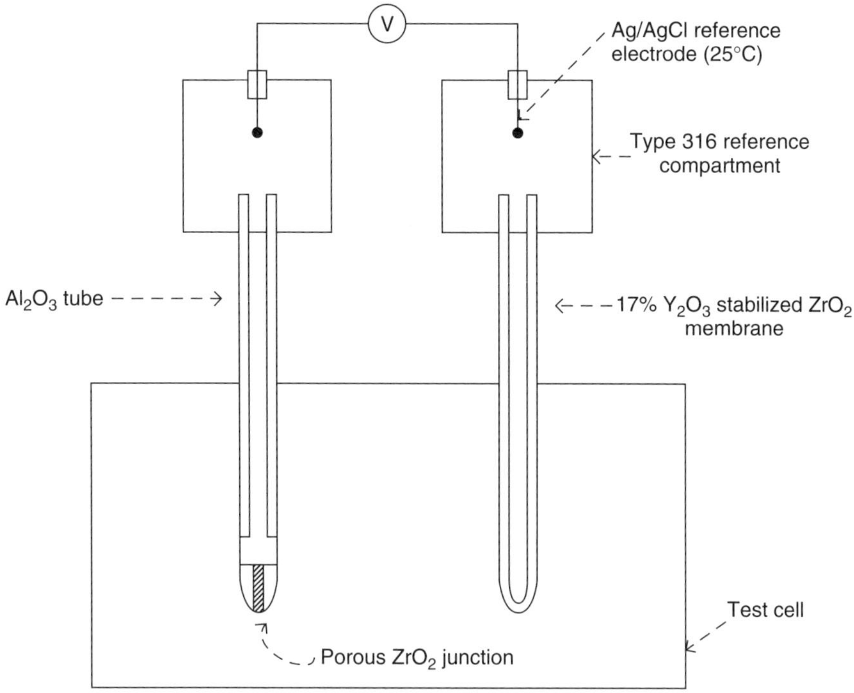

Fig. 1.20 Schematic of the ceramic membrane pH-measuring system (Eq. 1.26) [84].

By using an YSZME containing a buffered internal solution (0.1 m KCl + 0.01 m $B(OH)_3$ + 0.01 m KOH) and a matched EPBRE containing the identical solution (with a liquid junction of porous ZrO_2), and by equipping both with Ag/AgCl electroactive elements at ambient temperature (Fig. 1.20), it was postulated that the TLJPs could be matched and hence eliminated from the measured cell potential. The measured cell potential is then given by

$$\Delta E = E_D + \Delta\Phi_m \quad [1.27]$$

where E_D and $\Delta\Phi_m$ are the isothermal liquid junction potential of the external reference electrode and the membrane potential, respectively. E_D can be obtained via Henderson's equation and the estimates used to correct the measured cell potentials. A theoretical analysis shows that the membrane potential should exhibit a Nernstian pH response except for a small deviation due to the difference in the activity of water in the internal and test solutions. To test the concept, a platinized platinum/hydrogen electrode was added to the test cell containing different buffers as indicated in the caption to Fig. 1.21, in order to compare the results with the ceramic membrane

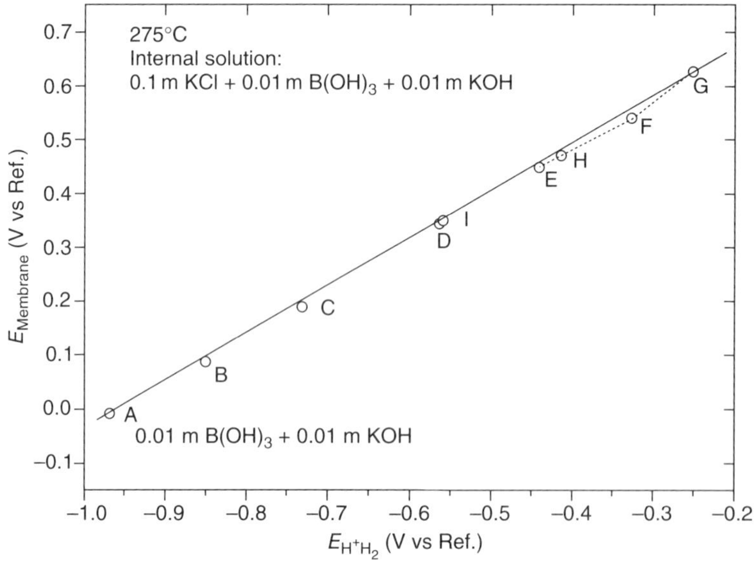

Fig. 1.21 Potential vs pH_T response for the ceramic membrane electrode (Eq. 1.26) in various $B(OH)_3$/KOH and H_2SO_4/Na_2SO_4 systems at 275°C [84]. The various test solutions are: (A) 0.01 m KOH + 0.01 m $B(OH)_3$, (B) 10^{-3} m KOH + 0.01 m $B(OH)_3$, (C) 10^{-4} m KOH + 0.01 m $B(OH)_3$, (D) 0.01 m $B(OH)_3$, (E) 0.00005 m H_2SO_4, (F) 0.0005 m H_2SO_4, (G) 0.005 m H_2SO_4, (H) 0.005 m H_2SO_4 + 0.005 m Na_2SO_4, (I) 0.005 m H_2SO_4 + 0.045 m Na_2SO_4.

electrode. Various test solutions were employed to calibrate the pH response of the ceramic membrane. The measured membrane potentials were plotted against the hydrogen electrode potentials at temperatures from 100 to 275°C, with Fig. 1.21 showing data at 275°C. It is indicated that the ZrO_2 (17% Y_2O_3) membrane does not exhibit a strictly Nernstian pH response, especially at the lower temperatures and in weakly acidic systems. Tsuruta and Macdonald have analyzed the possible reason, concluding that the rest of the error might be attributed to the effect of ionic strengths on the buffer properties of the internal solution and to errors in the estimates of E_D. Retrospectively, it is now believed that the use of a silica 'sintering aid' by the manufacturer of the YSZ ceramic tubes, resulting in abnormally high membrane impedance, is the principal cause of the deviations from theory at the lower temperatures. While this approach is clearly a most promising way of eliminating the TLJP in high temperature pH measurement, it has yet to be fully explored.

Macdonald and Kriksunov employed a YSZME electrode (internal element Ag/O_2) against the EPBRE to explore changes in acidity of dilute HCl and NaOH solutions for temperature ranging from 400 to 528°C as the

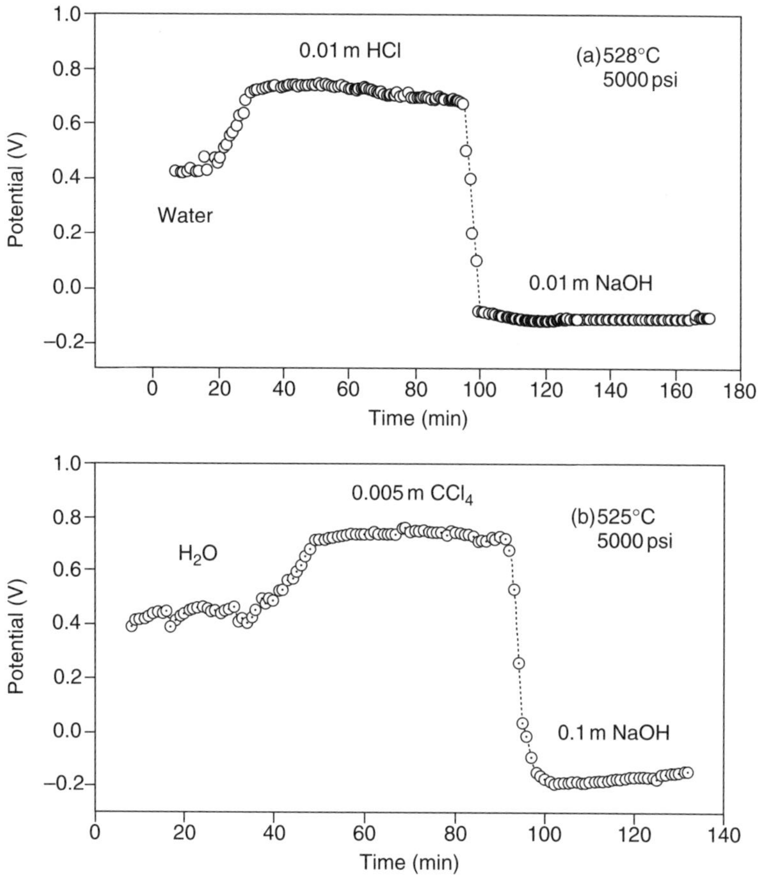

Fig. 1.22 Measured potential vs time for dilute HCl and NaOH solutions and for carbon tetrachloride in supercritical aqueous systems at temperature (a) 528°C and (b) 525°C [43].

solution cycled from 0.01 m HCl to 0.01 m NaOH, as shown in Fig. 1.22 [43]. The utility of the sensor for detecting pH changes in response to changes in solution composition at temperatures well above the critical point is clearly indicated from the figure. By using the calculated pH values and the measured cell potentials, Macdonald and Kriksunov calibrated the reference electrode and obtained the pH of CCl_4 solution assuming complete hydrolysis of CCl_4 (Fig. 1.23 [43]).

The pH of the CCl_4 solution was found to be a little higher than that calculated, assuming complete thermal hydrolysis of carbon tetrachloride. However, the level of agreement is still acceptable, considering the unknown actual extent of CCl_4 thermal hydrolysis and the uncertainty in the calibrating cell voltages. While the level of uncertainty perhaps does not satisfy the

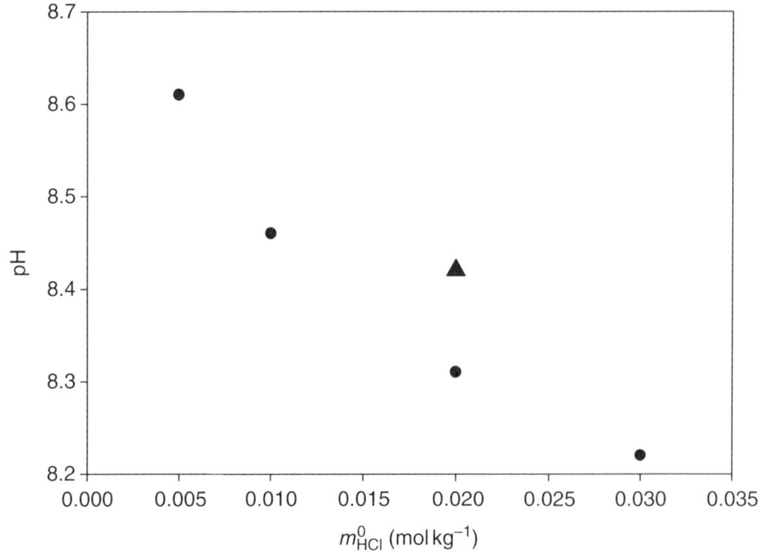

Fig. 1.23 Plot of calculated pH against stoichiometric HCl concentration for $T = 525°C$ and $p = 5000$ psi (closed circles). The measured pH for the hydrolyzed CCl_4 solution is given by the closed triangle [43].

need for accurate thermodynamic work, it is satisfactory for monitoring SCWO systems [31].

In the flow-through EPBREs (FTEPBREs) [50, 51] (Fig. 1.14), fluid flow was employed instead of pressure pulses to suppress thermal diffusion. This was done by pumping the internal reference solution through the inner compartment at a delicately set rate. The flow rate must be high enough to yield a constant electrolyte concentration along the nonisothermal bridge, but also not too high so as to maintain the temperature gradient along the electrolyte bridge. Since the concentration is kept constant along the bridge, the Soret initial state can be maintained and the thermal diffusion potential is constant for a given temperature difference across the junction. Thus, the uncertainty of the TLJP is assumed to be eliminated at any given temperature and pressure.

To test this assumption, Lvov *et al.* [51] employed two hydrochloride solutions of different concentrations (0.01 and 0.001 mol kg^{-1} HCl) as the test solutions, and measured the potential difference of the thermocell:

$$\begin{array}{l} |\leftarrow\ T_1\ \rightarrow|\leftarrow\qquad T_2 \qquad\rightarrow|\leftarrow\ T_1 \\ \text{Cu/Ag/AgCl/0.1 mol kg}^{-1}\text{ NaCl(aq.) | 0.1 mol kg}^{-1}\text{ NaCl(aq.) + HCl(aq.)/H}_2\text{,Pt/Cu} \end{array} \quad [1.28]$$

with the potential difference being:

$$\Delta E = E(10^{-2}\,\text{mol kg}^{-1}\text{HCl}) - E(10^{-3}\,\text{mol kg}^{-1}\text{HCl})$$
$$= \frac{RT_2}{F}\ln\frac{m_{\text{H}^+}(1)\gamma_{\text{H}^+}(1)}{m_{\text{H}^+}(2)\gamma_{\text{H}^+}(2)} + E_{\text{D}}[\text{HCl}(1)/\text{NaCl}]$$
$$- E_{\text{D}}[\text{HCl}(2)/\text{NaCl}]. \quad [1.29]$$

The potential difference for the two different solutions was shown to be independent of the thermal diffusion potential and can be calculated theoretically. The calculated and measured potentials, listed in Table 1.1, show good agreement and demonstrate the effectiveness of the flow-through technique in maintaining the Soret initial state. At the higher temperatures (350–360°C), the difference between the measured and calculated potentials was of the order of 10 mV, corresponding to an uncertainty in a measured pH value (if it was used for that purpose) of about 0.08 (±0.04).

Lvov *et al.* [52] used a flow-through electrochemical cell (comprising a flow-through external Ag/AgCl PBRE and a modified flow-through platinum hydrogen electrode) to measure pH at temperatures up to 400°C and pressures up to 25.3 MPa. The measured accuracy was demonstrated to be better than ±0.03 logarithmic units, in good agreement with the estimate calculated above.

A similar FTEPBRE has been employed by Lvov *et al.* [96] in a four-way, once-through electrochemical cell (composed of one flow-through reference electrode, one flow-through Pt-H_2 electrode, one flow-through YSZ electrode, and a thermocouple). Because the same dependence on the temperature, pressure, flow rate and solution component of the platinum electrode and YSZ electrode exist, the cell was run at 320 and 350°C (pressure

Table 1.1 Experimental and theoretical differences in potentials $E(10^{-2}$ mol kg^{-1} HCl) – $E(10^{-3}$ mol kg^{-1} HCl) of the thermocell (Eq. 1.28) as a function of temperature and pressure [3]

Temperature (K)	Pressure (bar)	*E*(0.01 mol kg^{-1} HCl) – *E*(0.001 mol kg^{-1} HCl)	
		Experimental observed (mV)	Theoretically calculated (mV)
298	275	64.5	59.1
298	338	61.4	59
373	275	81	74
473	275	97.8	93.7
573	275	100.3	109.8
623	275	93.4	102.2
623	338	97.4	106.2
633	338	91.7	102.4

23.0 and 24.8 MPa) to test the precision and response time of the YSZ electrode as a pH-sensing electrode.

Using a similar concept, Sue *et al.* [97] designed another flow-through EPBRE using a platinum/hydrogen electrode instead of an Ag/AgCl electrode. The cell they developed can be represented by:

$$\begin{array}{l} \mathrm{Cu/Pt,H_2/Test\ Solution \mid Reference\ Solution/H_2,Pt/Cu} \\ T_1 \rightarrow\mid\leftarrow \quad T_2 \quad \rightarrow\mid\leftarrow \quad T_1 \quad \rightarrow\mid \end{array} \qquad [1.30]$$

in which T_2 is the higher temperature and T_1 is the lower (ambient) temperature. The platinum/hydrogen electrode on the right side is the flow-through EPBRE while that on the left side is the indicator electrode. This electrochemical cell has been used to measure pH at temperatures from 23.9 to 400.2°C and pressures from 25.0 to 35.1 MPa. The difference between the measured pH and the calculated pH proved to be less than 0.03 logarithmic units. Sue *et al.* [98] also used the cell to measure the dissociation constant of phenol (PhOH) [98].

The development of FTEPBREs represents a significant advance in the quest to make possible accurate pH measurements in supercritical aqueous systems. However, there are also a number of issues that need to be resolved. One is the delicate setting of flow rate to simultaneously maintain the system in the Soret initial state and maintain the temperature gradient down the nonisothermal bridge, which greatly complicates the system and makes the FTEPBRE unlikely to be used in many practical field situations (e.g., SCWO system). The second is the calibration of the reference electrode. Since the reference potential (measured against a hydrogen electrode) strongly depends on the flow rate of the reference solution, it needs to be calibrated against a known standard over wide ranges of temperature, pressure, and solution type and composition, at accurately known flow rates. This feature alone indicates that the flow-through concept, which was initially introduced by Danielson [81] more than 20 years ago, is not the panacea in reference electrode technology.

1.6 Summary and conclusions

The current status of development of reference electrodes for use in high temperature/high pressure aqueous solutions has been reviewed, with emphasis on reference electrode technology for measuring potentials in high subcritical (300°C < T < 374.15°C) and in supercritical (T > 374.15°C) aqueous solutions. The review covers both internal and external reference electrodes and summarizes the advantages and disadvantages of both, particularly for use at supercritical temperatures. The design philosophies adopted in each type, particularly with regard to avoiding the irreversible contributions of the isothermal liquid junction and the thermal liquid junction to the measured potential, while avoiding the thermal hydrolysis of the electroactive element, are discussed in depth. It is concluded that no generally applicable reference electrode technology for measuring potentials in high subcritical and supercritical aqueous solutions has been developed, even though it remains the most pressing need in electrochemical studies in this field.

1.7 Acknowledgments

The authors gratefully acknowledge the support of this work by the US Army under Grant No. DAAD 190310040.

1.8 References

1. E. U. Franck. *ACS Symp. Series* **60**, 99 (1977).
2. W. M. Flarsheim, A. C. McDonald, A. J. Bard, K. P. Johnston. *Preprints of Papers – ACS, Div. Fuel Chem.* **30**, 16 (1985).
3. D. D. Macdonald, L. B. Kriksunov. *Electrochim. Acta* **47**, 775 (2001).
4. M. Hodes, P. A. Marrone, G. T. Hong, K. A. Smith, J. W. Tester. *J. Supercrit. Fluids* **29**, 265 (2004).
5. G. W. Morey. *Econ. Geol.* **52**, 225 (1957).
6. D. D. Macdonald, S. Hettiarachchi, S. J. Lenhart. *J. Soln Chem.* **17**, 719 (1988).
7. D. D. Macdonald, L. B. Kriksunov. *Corrosion* **34**, 75 (1978).
8. H. Toua, G. A. Mansoori. *Fluid Phase Equil.* **150–151**, 459 (1998).
9. R. E. Mesmer, C. F. Baes, Jr. *Inorg. Chem.* **10**, 2290 (1971).
10. R. E. Mesmer, C. F. Baes, Jr, F. H. Sweeton. *Inorg. Chem.* **11**, 537 (1972).
11. R. E. Mesmer, C. F. Baes, Jr. *J. Soln Chem.* **3**, 307 (1974).
12. F. H. Sweeton, R. E. Mesmer, C. F. Baes, Jr. *J. Phys. E:Sci. Instrum.* **6**, 165 (1973).
13. B. F. Hitch, R. E. Mesmer. *J. Soln Chem.* **5**, 667 (1976).
14. R. E. Mesmer, H. F. Holmes. *J. Soln Chem.* **21**, 725 (1992).
15. D. A. Palmer, D. J. Wesolowski. *Geochim. Cosmochim. Acta* **56**, 1093 (1992).
16. D. A. Palmer, D. J. Wesolowski. *Geochim. Cosmochim. Acta* **57**, 2929 (1993).
17. D. J. Wesolowski, D. A. Palmer, R. E. Mesmer. *Water–Rock Interaction: Proc. 8th Int. Symp. Water–Rock Interactions*, Vladivostok, Russia, 15–19 August, 1995 (ed. Y. K. Kharaka and O. V. Chudev), p. 51 (1995).
18. P. Benezeth, D. A. Palmer, D. J. Wesolowski. *Geothermics* **26**, 465 (1997).
19. P. C. Ho, D. A. Palmer, R. H. Wood. *J. Phys. Chem. B* **104**, 12084 (2000).
20. P. Benezeth, D. A. Palmer, D. J. Wesolowski. *J. Chem. Engng Data* **46**, 202 (2001).
21. P. Benezeth, D. J. Wesolowski, D. A. Palmer. *J. Chem. Engng Data* **48**, 171 (2003).
22. D. D. Macdonald, P. Butler, D. Owen. *Can. J. Chem.* **51**, 2590 (1973).
23. D. D. Macdonald, P. Butler, D. Owen. *J. Phys. Chem.* **77**, 2474 (1973).
24. D. D. Macdonald, D. Owen. *Can. J. Chem.* **51**, 2747 (1973).
25. D. W. Shoesmith, L. Woon. *Can. J. Chem.* **54**, 3553 (1976).

26. G. Giasson, P. H. Tewari. *Can. J. Chem.* **56**, 435 (1978).
27. Y. Matsushima, A. Matsunaga, K. Sakai, A. Okuwaki. *Bull. Chem. Soc. Japan* **61**, 4259 (1988).
28. B. A. Bilal, E. Mueller. *Z. Naturforsch.* **48A**, 1073 (1993).
29. L. W. Niedrach. *J. Electrochem. Soc.* **127**, 2122 (1980).
30. T. Tsuruta, D. D. Macdonald. *J. Electrochem. Soc.* **129**, 1202 (1982).
31. M. J. Danielson, O. H. Koski, J. Meyers. *J. Electrochem. Soc.* **132**, 296 (1985).
32. M. J. Danielson, O. H. Koski, J. Meyers. *J. Electrochem. Soc.* **132**, 2037 (1985).
33. W. L. Bourcier, G. C. Ulmer, H. L. Barnes. *Hydrothermal Experimental Techniques*, p. 157 (Wiley-Interscience, New York, 1987).
34. L. W. Niedrach. *Science* **207**, 1200 (1980).
35. L. W. Niedrach. Patent Number, 4264424 (General Electric, 1981).
36. L. W. Niedrach. *J. Electrochem. Soc.* **212**, 672 (1982).
37. L. W. Niedrach. *Adv. Ceram. Sci.* **12**, 672 (1984).
38. S. Hettiarachchi, D. D. Macdonald. *J. Electrochem. Soc.* **131**, 2206 (1984).
39. D. D. Macdonald, S. Hettiarachchi, H. Song, K. Makela, R. Emerson, M. Haim. *J. Soln Chem.* **21**, 849 (1992).
40. S. Hettiarachchi, K. Makela, H. Song, D. D. Macdonald. *J. Electrochem. Soc.* **139**, L3–L4 (1992).
41. L. B. Kriksunov, D. D. Macdonald. *ASME Heat Transfer Div Publ HTD, v 317-2, ASME Heat Transfer Division* 271 (1995).
42. S. N. Lvov, G. Perboni, M. Broglia. *Physical Chemistry of Aqueous Systems*, p. 441 (Begell House, New York, 1995).
43. D. D. Macdonald, L. B. Kriksunov. In *Proc. Symp. Emerging Technologies in Hazardous Waste Management VII* (ed. D. W. Tedder) pp. 1280–3 (American Chemical Society, Washington, DC, 1995).
44. L. B. Kriksunov, D. D. Macdonald. In *Proc. Symp. High Temp. Press. Soln Chem., Int. Conf. Pacific Basin Chem. Socs, PACIFICHEM '95* (Washington, DC, 1995).
45. K. Ding, W. E. Seyfried, Jr. *Geochim. Cosmochim. Acta* **59**, 4769 (1995).
46. K. Eklund, S. N. Lvov, D. D. Macdonald. *J. Electroanalyt. Chem.* **437**, 99 (1997).
47. R. Fernandez-Prini, R. Crovetto. *J. Phys. Chem. Ref. Data* **18**, 1231 (1989).
48. N. Kishima, H. Sakai. *Earth Planetary Sci. Lett.* **67**, 79 (1984).
49. K. Ding, W. E. Seyfried, Jr. *J. Soln Chem.* **25**, 421 (1996).
50. S. N. Lvov, H. Gao, D. D. Macdonald. in *Proc. Fifth Int. Symp. Hydrothermal Reactions.* (ed. D. A. Palmer and D. Wesolowski) 146 (ORNL, US Dept of Energy, Gatlinburg, TN, 1997).
51. S. N. Lvov, H. Gao, D. D. Macdonald. *J. Electroanalyt. Chem.* **443**, 186 (1998).
52. S. N. Lvov, X. Y. Zhou, D. D. Macdonald. *J. Electroanalyt. Chem.* **463**, 146 (1999).
53. K. Sue, M. Uchida, T. Usami, T. Adschiri, K. Arai. *J. Supercrit. Fluids* **28**, 287 (2004).
54. K. Sue, M. Uchida, T. Adschiri, K. Arai. *J. Supercrit. Fluids* **31**, 295 (2004)
55. M. H. Lietzke. *J. Am. Chem. Soc.* **77**, 1344 (1955).
56. R. S. Greeley, W. T. Smith, R. W. Stoughton, M. H. Lietzke. *J. Phys. Chem.* **64**, 652 (1960).
57. R. S. Greeley, W. T. Smith, M. H. Lietzke, R. W. Stoughton. *J. Phys. Chem.* **64**, 1445 (1960).

58. M. H. Lietzke, H. B. Hupf, R. W. Stoughton. *J. Phys. Chem.* **69**, 2395 (1965).
59. B. Case, G. J. Bignold. *J. Appl. Electrochem.* **1**, 141 (1971).
60. K. Ding, W. E. Seyfried, Jr. *Science* **272**, 1634 (1996).
61. G. B. Naumov, B. N. Ryzhenko, I. L. Khodakovsky. *Handbook of Thermodynamic Data* (ed. I. Barnes and V. Speltz) (US Geological Survey, Menlo Park, CA, 1974).
62. J. C. Tanger IV, H. C. Helgeson. *Am. J. Sci.* **288**, 19 (1988).
63. D. A. Sverjensky, J. J. Hemley, W. M. D'Angelo. *Geochim. Cosmochim. Acta* **55**, 988 (1991).
64. J. W. Johnson, E. H. Oelkers, H. C. Helgeson. *Comput. Geosci.* **18**, 899 (1992).
65. E. L. Shock, E. H. Oelkers, J. W. Johnson, D. A. Sverjensky, H. C. Helgeson. *J. Chem. Soc., Faraday Trans.* **88**, 803 (1992).
66. K. Ding, W. E. Seyfried, Jr. *Geochim. Cosmochim. Acta* **56**, 3681 (1992).
67. J. V. Dobson, M. N. Dagless, H. R. Thirsk. *J. Chem. Soc., Faraday Trans.* 1 **68**, 739 (1972).
68. J. V. Dobson, M. N. Dagless, H. R. Thirsk. *J. Chem. Soc., Faraday Trans.* 1 **68**, 764 (1972).
69. J. V. Dobson, B. R. Chapman, H. R. Thirsk. In *Int. Conf. High Temperature High Pressure Electrochemistry in Aqueous Solutions*, NACE International, Houston, TX (1973).
70. D. D. Macdonald. In *Modern Aspects of Electrochemistry* (ed. J. O'M. Bockris and B. E. Conway), p. 141 (Plenum Press, New York, 1975).
71. D. D. Macdonald, P. R. Wentrcek, A. C. Scott. *J. Electrochem. Soc.* **127**, 1745 (1980).
72. Z. Nagy, R. M. Yonco. *J. Electrochem. Soc.* **133**, 2232 (1986).
73. S. Hettiarachchi, D. D. Macdonald. *J. Electrochem. Soc.* **134**, 1307 (1987).
74. P. Jayaweera, T. O. Passell, P. J. Millett. Patent Number, US 5425871.
75. D. de Jones, H. C. Masterson. *Adv. Corr. Sci. Tech.* **1**, 1 (1970).
76. D. D. Macdonald, D. Owen. *J. Electrochem. Soc.* **120**, 317 (1973).
77. D. D. Macdonald, A. C. Scott, P. Wentrcek. *J. Electrochem. Soc.: Electrochem. Sci. Technol.* **126**, 908 (1979).
78. D. D. Macdonald, S. Hettiarachchi, S. J. Lenhart. In *Proc. 1987 Symp. Chemistry in High Temperature Water (EPRI Report NP-6005, 1990)* (Brigham Young University, Provo, UT, 1990).
79. S. Hettiarachchi, S. J. Lenhart, D. D. Macdonald. In *Proc. 3rd Int. Symp. Environ. Degrad. Mat. Nucl. Power Systs – Water Reactors* (ed. G. J. Theus and J. R. Weeks), p. 165 (Met. Soc. AIME, Warrendale, PA, 1988).
80. D. D. Macdonald, S. Hettiarachchi, R. Emerson, K. Makela, H. Song, M. B. Haim. In *Proc. Symp. Chemistry in High-Temperature Aqueous Solutions* C3a-1 (Brigham Young University, Provo, UT, 1991).
81. M. J. Danielson. *Corrosion* **39**, 202 (1983).
82. S. N. Lvov, X. Y. Zhou. *Mineralogical Magazine* **62A**, 929 (1998).
83. A. J. Bard, L. R. Faulkner. In *Electrochemical Methods Fundamentals and Applications,* 2nd edn, p. 63 (Wiley, New York, 2000).
84. T. Tsuruta, D. D. Macdonald. *J. Electrochem. Soc.* **129**, 1221 (1982).
85. H. W. Harper. *J. Phys. Chem.* **89**, 1659 (1985).
86. R. Haase. *Thermodynamics of Irreversible Processes* (Addison-Wesley, Reading, MA, 1969).

87. S. N. Lvov, D. D. Macdonald. *J. Electroanalyt. Chem.* **403**, 25 (1996).
88. G. R. Engelhardt, S. N. Lvov, D. D. Macdonald. *J. Electroanalyt. Chem.* **429**, 193 (1997).
89. D. D. Macdonald, A. C. Scott, P. R. Wentrcek. *J. Electrochem. Soc.* **126**, 1618 (1979).
90. L. S. Hwang, A. Boateng, D. D. Macdonald. In *Proc. Symp. Corrosion in Batteries and Fuel Cells and Corrosion Solar Energy Systems.* (ed. C. J. Johnson and S. L. Pohlman), p. 492 (Electrochemical Society, Inc., Pennington, NJ, 1983).
91. N. Akinfiev, S. N. Lvov, D. D. Macdonald. *Comput. Geosci.* (in press).
92. W. J. Johnson, E. H. Qelkers, H. C. Helgeson. *Comput. Geosci.* **18**, 899 (1992).
93. A. S. Quist, W. L. Marshall. *J. Phys. Chem.* **69**, 2984 (1965).
94. Y. D. Rakhmilevich, I. A. Dibrov, S. N. Lvov. *Russ. J. Phys. Chem.* **61**, 2391 (1987).
95. S. N. Lvov, A. Marcomini, M. M. Suprun. In *Proc. 4th Int. Symp. Hydrothermal Reactions* (ed. M. Cuney and M. Cathelineau), p. 135 (Nancy, 1993).
96. S. N. Lvov, X. Y. Zhou, G. C. Ulmer, H. L. Barnes, D. D. Macdonald, S. M. Ulyanov, L. G. Benning, D. E. Grandstaff, M. Manna, E. Vicenzi. *Chem. Geol.* **198**, 141 (2003).
97. K. Sue, K. Murata, Y. Matsuura, M. Tsukagoshi, T. Adschiri, K. Arai. *Rev. Sci. Instrum.* **72**, 4442 (2001).
98. K. Sue, K. Murata, Y. Matsuura, M. Tsukagoshi, T. Adschiri, K. Arai. *Fluid Phase Equilib.* **194–197**, 1097 (2002).

2
LIRES: a European-sponsored research project to develop light water reactor reference electrodes

R.-W. BOSCH, SCK·CEN, Belgium; Z. KERNER and G. NAGY, AEKI, Hungary; D. FÉRON, CEA, France; M. NAVAS, CIEMAT, Spain; W. BOGAERTS, KU Leuven, Belgium; D. KÁRNÍK, NRI Rez, Czech Republic; T. DORSCH and R. KILIAN, Framatome ANP, Germany; M. ULLBERG and A. MOLANDER, Studsvik Nuclear AB, Sweden; K. MÄKELÄ and T. SAARIO, VTT Industrial Systems, Finland

2.1 Introduction

With the increasing age of nuclear power plants, stainless steel core components suffer from increasing irradiation damage. Cracking in such core components occurs due to a combination of high temperature, stress and irradiation, and is strongly affected by the corrosion potential. The corrosion potential is a useful measure for monitoring of cracking as it can help to identify when it is likely to occur (high value of the corrosion potential) and when it is not likely to occur (low value of the corrosion potential). For example, irradiation-assisted stress corrosion cracking (IASCC) has been observed in both boiling water reactors (BWRs) and pressurized water reactors (PWRs) [1, 2]. The corrosion potential is used for monitoring and mitigating IASCC, since IASCC of stainless steels in BWRs can be prevented if the corrosion potential is decreased below –230 mV (standard hydrogen electrode, SHE) [3]. IASCC occurs above a threshold fast neutron dose which depends on the reactor type (BWR vs PWR). To measure the corrosion potential a reference electrode is needed.

The main objective of the LIRES project is to develop reference electrodes that can be used inside the core of a light water reactor (LWR) (hence LIRES: light water reference electrodes). Thus the electrodes must survive under high temperature, pressure and irradiation conditions, for a sufficiently long life time, i.e., at least one fuel cycle. Also the potential of the reference electrode must be related to the SHE scale. One approach to such corrosion potential measurements is shown in Fig. 2.1.

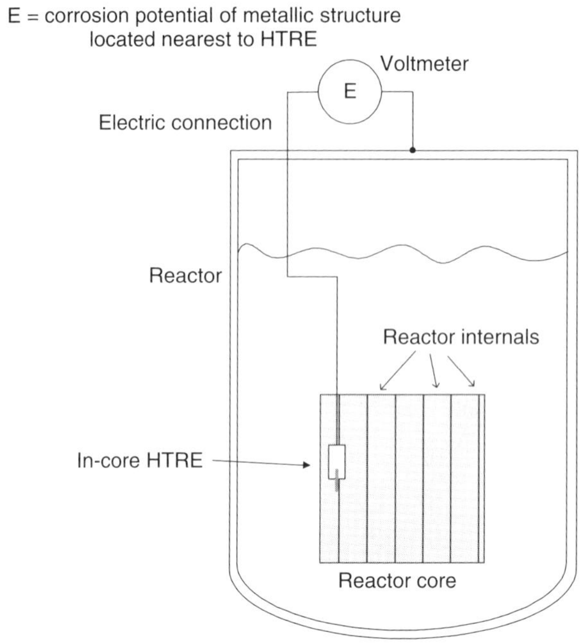

Fig. 2.1 Schematic of an in-core high temperature reference electrode measuring a corrosion potential of a reactor internal. HTRE, high temperature reference electrode.

Besides the measurement of the corrosion potential, other applications are possible such as the measurement of the oxidation–reduction potential with a platinum electrode (which is a measure of the oxidizing characteristics of the solution) and pH measurements if there is a suitable pH electrode.

A four-step development trajectory is foreseen:

1 Definition of a testing standard for high temperature reference electrodes to be used during the round robin test. This testing standard should be specific and clear enough to ensure reliable and reproducible results when used by different laboratories.
2 Design and development of high temperature reference electrodes for PWR and BWR conditions with operating temperature up to 350°C at laboratory scale.
3 Round robin testing among the participating laboratories of the new developed reference electrodes, using the test procedure developed under step 1. Based on the round robin results, the best reference electrodes will be selected and used for the irradiation experiment.
4 Testing of high temperature reference electrodes under appropriate irradiation conditions in a material test reactor. One irradiation will be

performed under BWR conditions and one irradiation will be performed under PWR conditions. Both high temperature reference electrodes (for BWR and PWR application) are to be selected from the round robin test.

Based on a literature review, regarding all kinds of high temperature reference electrodes, the reference electrodes were categorized [4, 5]. Six categories of reference electrodes were distinguished:

1 *External pressure-balanced reference electrode*
The external pressure-balanced reference electrodes (EPBREs) offer considerable benefits in terms of long-term stability and serviceability. They are suitable to measure potentials when moderate accuracy (±25 mV) is acceptable (e.g., in corrosion systems). EPBREs can be calibrated to the standard hydrogen electrode (SHE) scale.

2 *Internal Ag/AgCl reference electrode*
To avoid the mixed potential produced by the presence of hydrogen, an Ag/AgCl internal reference electrode can be used, which contains high purity water enclosed in a sapphire body. The electrode potential is determined by the solubility of silver chloride in high purity water at temperatures up to 275°C. This electrode has been operated successfully in nuclear power plants with hydrogen water chemistry not exceeding 200 ppb dissolved hydrogen. However, the lifetime of such an electrode is limited due to degradation of the AgCl layer by the high temperature and presence of hydrogen.

3 *Inorganic polymer electrolyte (sodium silicate) reference electrodes*
An inorganic polymer electrolyte seems to be a promising design, consisting of polymerized sodium silicate (highly gelled form) and doped with high amounts of Cl^-. This electrode exhibits a fast and stable response up to 300°C. However, prediction of the electrode potential is complex because the activity of Cl^- and the activity coefficient of KCl in a solid electrolyte matrix are not well known.

4 *Yttrium-stabilized zirconium oxide reference electrode*
Yttrium-stabilized zirconia (YSZ) membrane and tungsten/tungsten oxide electrodes show a reversible response to pH at temperatures up to 300°C. These electrodes therefore can be used as reference electrodes if used at a known pH.

5 *Palladium/hydride, Pt and other noble metal reference electrodes*
Electrodes of the palladium/hydride type are reversible with pH and can also operate as reference electrodes in a solution of constant pH. When the palladium/hydride electrode is utilized at temperatures higher than 275°C, caution is recommended due to deviations from predicted Nernstian responses.

6 *New reference electrode design concepts*
New reference electrode design concepts have emerged from the use of kinetics of the hydrogen reaction on palladium and platinum type substrates.

Based on experience available in the research consortium and a literature review, the following high temperature reference electrodes were selected to be further developed [5, 6]: a YSZ nickel/nickel oxide electrode, an external Ag/AgCl electrode, a rhodium electrode and a palladium/hydride electrode.

2.2 Testing standard

A testing standard has been defined that can be used to test the proper functioning of reference electrodes in water at elevated temperatures. The procedures and methodologies described in this standard should be specific and clear enough to ensure reliable and reproducible results when used by different laboratories. The testing standard is based on the flow chart shown in Fig. 2.2.

Distinction was made between testing at room temperature and testing at high temperature. After a measurement equipment check-up (step 2.1) tests are to be performed at room temperature, consisting of a pressure test, an electrical insulation test, a comparison test with a commercial reference

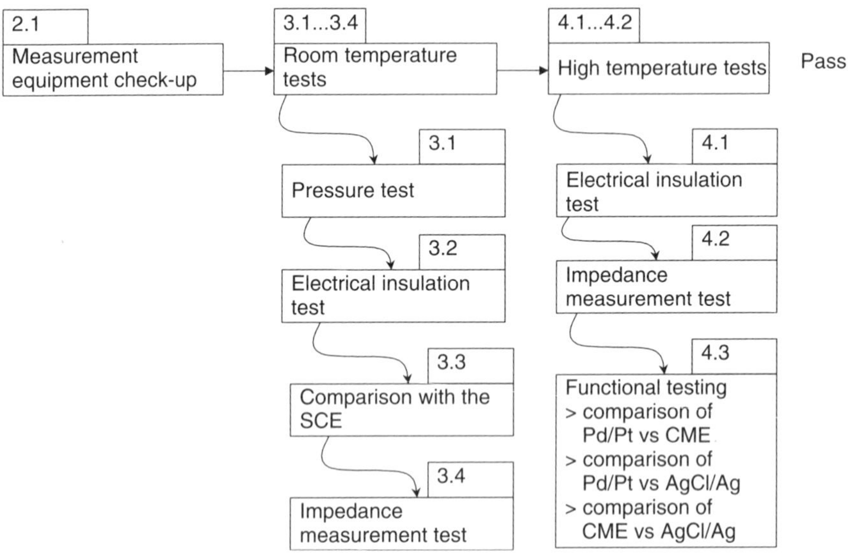

Fig. 2.2 Testing scheme for reference electrodes in high temperature water. SCE, saturated calomel electrode; CME, ceramic membrane electrode.

Table 2.1 Potentials to be measured between the different reference electrodes

No.	Electrode I	Electrode II
1	YSZ electrode	Pt wire electrode
2	External Ag/AgCl electrode	Pt wire electrode
3	Rh electrode	Pt wire electrode
4	Pd electrode	Pt wire electrode

electrode and an impedance test of the reference electrode system. The latter is necessary because when the impedance or resistance of the reference electrode system exceeds 1 MΩ, the potential measurement becomes complicated and too sensitive to noise. Then testing at high temperature will be performed, consisting of an electrical insulation test, an impedance measurement test and functional testing against other high temperature reference electrodes (like EPBREs or Pt electrodes).

The functional testing consists of the measurement of the potential difference of the new reference electrode against well-known reference electrodes. It was decided that all the potentials of the reference electrodes will be compared against each other including a simple Pt wire electrode. Measurements will be performed under PWR conditions (1200 ppm B and 2 ppm Li) at three different temperatures: 300, 325 and 350°C. It is assumed that when the electrode survives mechanically and chemically under PWR conditions, the reference electrode can also be used under BWR conditions. For the test at 350°C, a Pt wire electrode with an oxidized Zr wire as signal cable is used. Table 2.1 summarizes the potentials to be measured.

2.3 High temperature reference electrodes

2.3.1 Mixed metal oxide reference electrode

The principle of the mixed metal oxide reference electrode is based on the zirconia pH sensor developed by Niedrach [7, 8]. This zirconia pH sensor can be considered as the high temperature analogue of the low temperature glass pH electrode. The mixed metal oxide reference electrode consists of an YSZ tube. The lower end of the tube is filled with a mixture of Ni/NiO powder. The rest of the tube is filled with aluminum oxide tubes to close the tube and isolate the metal wire, which connects the Ni/NiO mixture with the mineral-insulated cable. The YSZ tube becomes an oxygen ion conductor at a sufficiently high temperature ($T > 180$°C), which makes it possible to use this tube as part of a reference electrode. The potential of the electrode is determined by the nickel–nickel oxide reference couple on the inside of the tube and the pH of the water on the outside of the tube [9]. A schematic and photograph of the reference electrode is shown in Fig. 2.3. This electrode is also frequently referred to as the ceramic membrane electrode (CME).

The most innovative part of the electrode design is the metal ceramic connection made by means of the fast magnetic forming (compression) technique. A nickel tube was connected to the ceramic tube by means of this magnetic compression technique. A strong magnetic field pulse compresses the nickel tube to the zirconia tube. A graphite layer is placed in between the ceramic and nickel tube to enhance a uniform force distribution. Figure 2.4 shows a cross-section of such a sealing.

This makes it possible to construct the whole high temperature reference electrode (HTRE) from radiation and high temperature resistant materials, resulting in a design that can survive inside an LWR under high temperature, high pressure and radiation. Test results under typical PWR conditions, a pressure of 150 bar and a temperature between 300 and 350°C, showed that the reference electrode is mechanically stable and gives reasonably reliable potential measurements.

The potential of the YSZ reference electrode is only dependent on the pH. That means that when the pH is known the YSZ electrode can be used

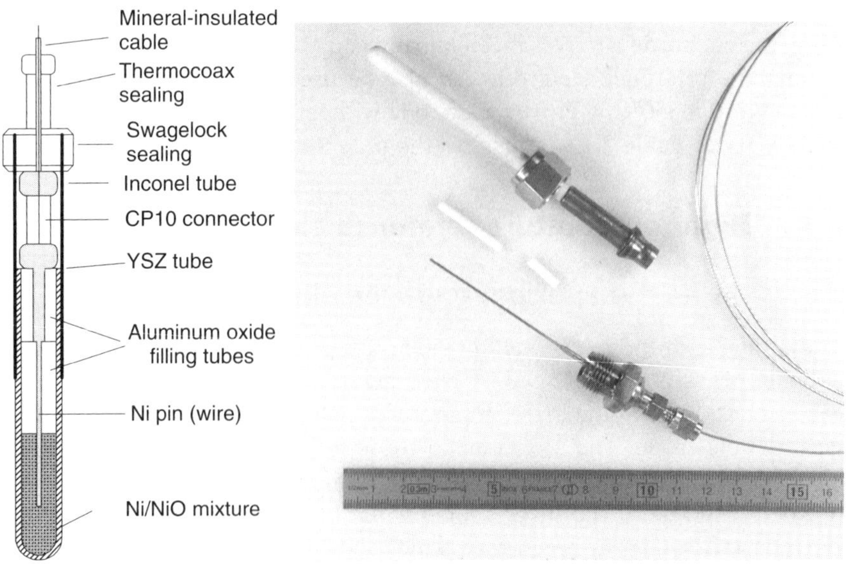

Fig. 2.3 Schematic of the yttrium-stabilized zirconia (YSZ) reference electrode and photograph of the final design.

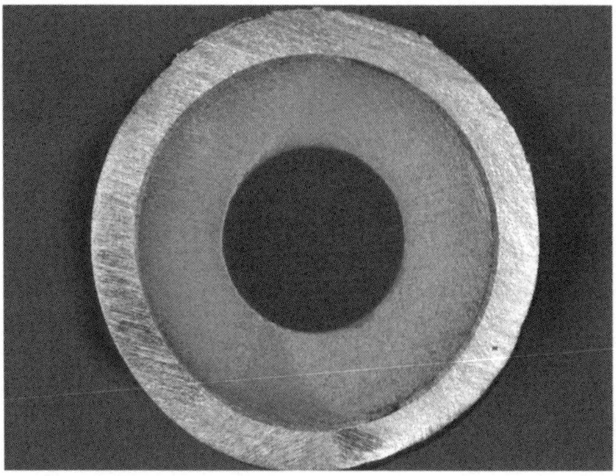

Fig. 2.4 Cross-section of a nickel–graphite–zirconium oxide sealing.

as a reference electrode. On the other hand when the pH is not known and a proper reference electrode is available, this electrode can be used to determine the pH of an aqueous solution at high temperature.

The electrode reaction that determines the electrode potential is [9]:

$$Ni + H_2O \Leftrightarrow NiO + 2H^+ + 2e \qquad [2.1]$$

The Nernst equation will give the electrode potential related to the SHE:

$$E = E^0_{Ni/NiO} - \frac{2.303RT}{F}pH \qquad [2.2]$$

where $E^0_{Ni/NiO}$ is the standard electrode potential for the reaction

$$NiO + H_2 \Leftrightarrow Ni + H_2O. \qquad [2.3]$$

The standard electrode potential can be calculated with the chemical potentials of the individual species taking part in the electrochemical reaction. Figure 2.5 shows the theoretically calculated standard electrode potential as a function of temperature with a linear and polynomial curve fit.

For a neutral solution the YSZ(Ni/NiO) potential as a function of temperature is shown in Fig. 2.6, including experimental data from VTT [10].

For a PWR solution with 1200 ppm boric acid and 2 ppm lithium hydroxide, the pH at 300°C is 6.9. At 300°C this results in a potential of −746 mV SHE for the YSZ(Ni/NiO) electrode. Table 2.2 shows a summary of YSZ electrode potentials as a function of pH and temperature. The potentials were calculated using the polynomial relationship for $E^0_{Ni/NiO}$ of Fig. 2.5.

2.3.2 External Ag/AgCl reference electrode

The Ag/AgCl reference electrode is probably one of the most used reference electrodes [11–15]. Unfortunately the Ag/AgCl redox couple cannot

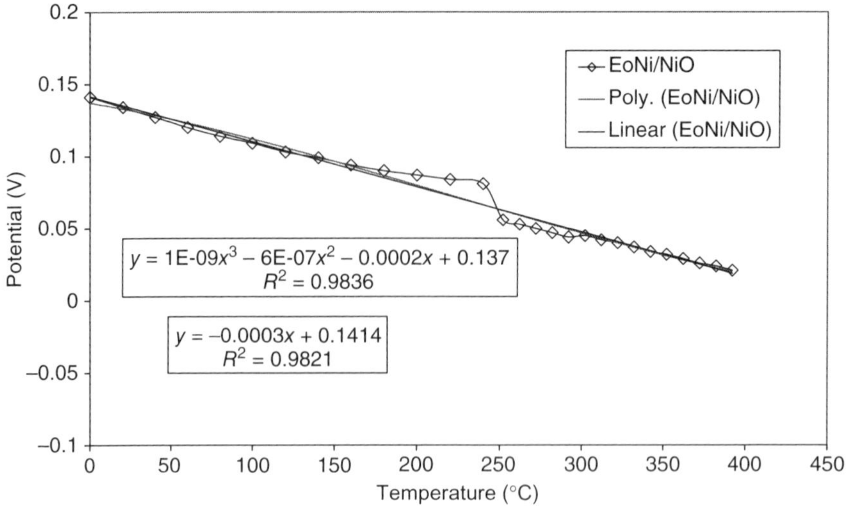

Fig. 2.5 Standard electrode potential of the yttrium-stabilized zirconia (Ni/NiO) electrode as a function of temperature.

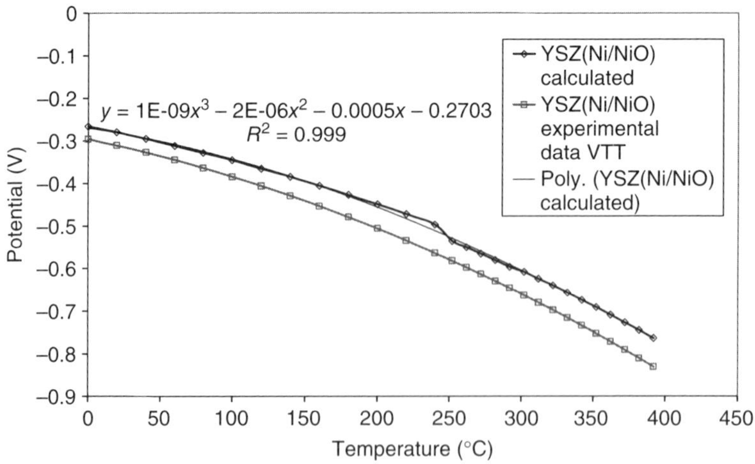

Fig. 2.6 Yttrium-stabilized zirconia [YSZ(Ni/NiO)] electrode potential for a neutral solution as a function of temperature.

be used at high temperatures due to the instability of the AgCl layer. Also this AgCl layer is sensitive to degradation by hydrogen when this gas is present in large quantities such as in BWRs under hydrogen water chemistry (HWC) and in PWRs. Therefore an external reference electrode based on the Ag/AgCl reference electrode couple was developed, which was connected to the test solution by means of a salt bridge. This salt bridge is then

Table 2.2 Yttrium-stabilized zirconium oxide reference electrode potentials under pressurized water reactor conditions

T (°C)	$E^0_{Ni/NiO}$	pH = 6.8	pH = 6.9	pH = 7.0	pH = 7.1	pH = 7.2	pH = 7.3
200	0.081	0.557	0.567	0.576	0.585	0.595	0.604
220	0.075	0.591	0.600	0.610	0.620	0.630	0.640
240	0.068	0.624	0.634	0.644	0.655	0.665	0.675
260	0.062	0.657	0.668	0.678	0.689	0.700	0.710
280	0.056	0.690	0.701	0.712	0.723	0.734	0.745
300	0.050	0.723	0.735	0.746	0.757	0.769	0.780
320	0.044	0.756	0.768	0.779	0.791	0.803	0.815
340	0.039	0.788	0.800	0.813	0.825	0.837	0.849
360	0.034	0.820	0.833	0.845	0.858	0.871	0.883

responsible for the transportation of the electrochemical information from the in-core location to the external reference electrode. The length of the salt bridge then determines to what extent the desired in-core location for corrosion potential measurements can be reached.

The external reference electrode consists of a sapphire tube (Al_2O_3 monocrystal) with an asbestos wick. Sapphire tubes were used as materials for the high temperature parts because of the stability and lifetime in boric acid environments.

An Ag wire with a diameter of 0.5 mm with an electrochemically deposited AgCl layer is placed inside the tube. This tube is filled with distilled water with a chloride concentration based on the maximum solubility of AgCl. The sapphire tube with the electrode system is placed in a stainless steel tube with an outer diameter of 6 mm. There is a water cooler for constant temperature maintenance of the electrode system. A connector head of the electrode with a BNC connector (central conductor is connected to the electrode system, shielding is connected to the body of electrode) serves for electrical connection of the electrode. The electrode can be connected to an autoclave via a Swagelok type coupling. Figures 2.7 and 2.8 show a schematic and a photograph, respectively, of the external reference electrode.

The selected structural materials enable uninterrupted service for a long time under primary circuit PWR conditions. In the case of mechanical damage of the sapphire tube or the electrode system, the electrode is quite safe from the point of view of pressure leakage.

The external electrode has been used for measuring corrosion potentials in the primary circuit under PWR conditions and has been used for the measurement of potentiodynamic polarization curves of primary circuit

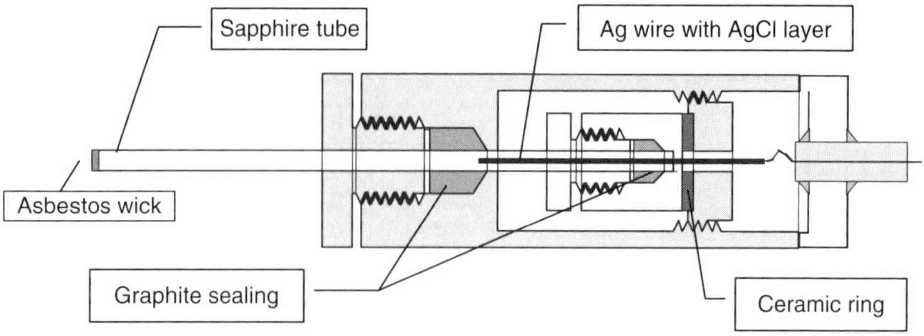

Fig. 2.7 Schematic of the external Ag/AgCl reference electrode.

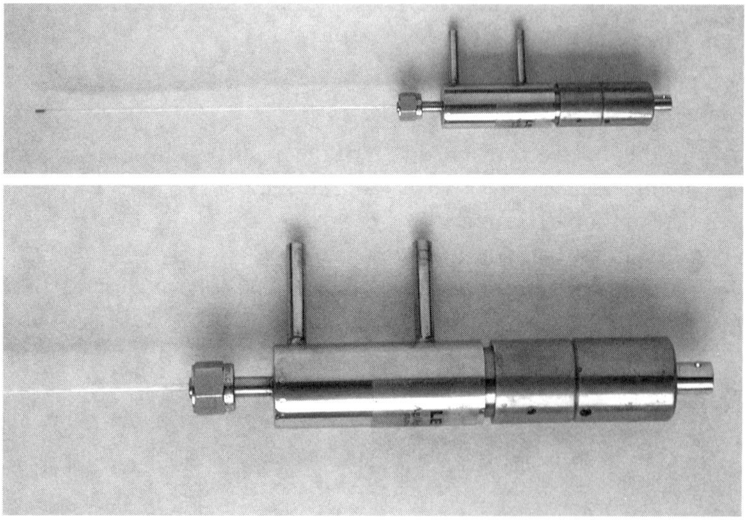

Fig. 2.8 External reference electrode Ag/AgCl.

materials in environments with crevices (strongly saturated salt solution) with good results.

The electrode has been calibrated against the standard hydrogen electrode, resulting in the following equation,

$$E_{SHE} = E_{meas} + E_{25} + 32.357 - 0.4299T - 1.1126\cdot 10^{-2}T^2 + 4.8983\cdot 10^{-5}T^3 - 7.8717\cdot 10^{-8}T^4 \qquad [2.4]$$

where T is the temperature in °C, E_{25} is the electrode potential at 25°C (determined by the solubility product of AgCl at 25°C), E_{meas} is the measured electrochemical potential against this reference electrode and E_{SHE} is the measured electrochemical potential against the SHE scale. Figure 2.9 shows a comparison between the thermal potential drops obtained from different sources.

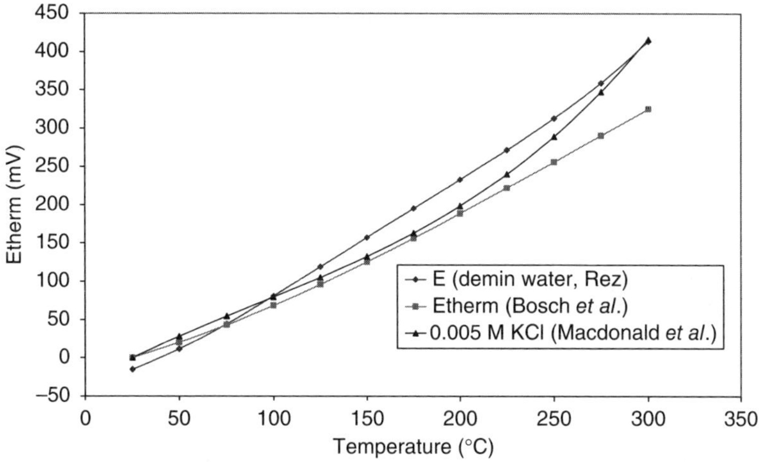

Fig. 2.9 Thermal potential drop from different sources: Rez (this chapter), Macdonald *et al.* [14] and Bosch *et al.* [15].

2.3.3 Rhodium electrode

The rhodium electrode has been developed with the emphasis on BWR application and more specifically for the monitoring of HWC. The main requirement of the electrochemical properties of the electrode is the capability to detect a departure from reducing conditions. In practice, this means that the electrode should be able to monitor weakly oxidizing conditions. More strongly oxidizing conditions are, however, of limited interest in connection with HWC supervision. A platinum hydrogen electrode is therefore often a preferred choice for HWC monitoring, mainly because of its excellent stability and long service life. Platinum electrodes have been used for more than 10 years in external pump BWRs, where the electrode can be located in a fitting in a pipe wall. A reliable Pt electrode for the difficult operating conditions inside the reactor pressure vessel, and in-core, would be a valuable addition to the materials-monitoring equipment available to BWR operators.

Most common is to use sapphire as insulator. Sapphire can be brazed to a metal, which is considered an advantage. However, the sapphire has been found to dissolve slowly in both BWR and PWR reactor water and the brazing can corrode and cause leakage. A more robust solution is therefore warranted. One of the most stable insulating materials available is zirconium oxide. The stability of this material towards reactor water is well demonstrated by the reliable performance of Zircaloy fuel cladding in water-cooled reactors. Zirconium oxide should therefore be a suitable insulating material to use in an in-core electrode.

The type of seals used in demanding applications in reactor technology is normally metallic C-ring seals or similar designs. A characteristic of metal seals of this type is that strong compressive forces are required for sealing. Platinum is a soft metal and is not suitable for use in combination with metal C-ring seals. Therefore platinum suitably alloyed with the harder noble metal rhodium has been used as the electrode material. A photograph of the rhodium electrode is shown in Fig. 2.10. Notice that the tip of the electrode is made from Rh/Pt alloy, which is not connected to the stainless steel part after the yellow/beige-colored ceramic insulation.

The question remains whether this Rh/Pt alloy shows the same electrochemical behavior as pure platinum. Therefore tests were carried out in an autoclave to investigate the electrochemical response of Rh to hydrogen and oxygen and to compare to the Pt response. The potentials of the Rh and Pt electrode were measured against two Ag/AgCl reference electrodes. Different combinations of hydrogen and oxygen concentrations were tested, corresponding to reducing, weakly oxidizing and oxidizing conditions. The results are summarized in Fig. 2.11.

The potentials of Rh and Pt respond similarly to variations in water chemistry. Also the values of the potentials are in the same region for similar water chemistries, showing that Rh is a good alternative for Pt. The electrode potential of the Rh electrode is assumed to be similar to the Pt electrode and can be calculated with the Nernst equation for the hydrogen equilibrium reaction $H_2 \Leftrightarrow 2H + 2e^-$ [16]. Figure 2.12 shows the hydrogen electrode potential for various hydrogen concentrations and temperatures. Figure 2.13 shows the hydrogen electrode potential for various pH values with a fixed hydrogen concentration.

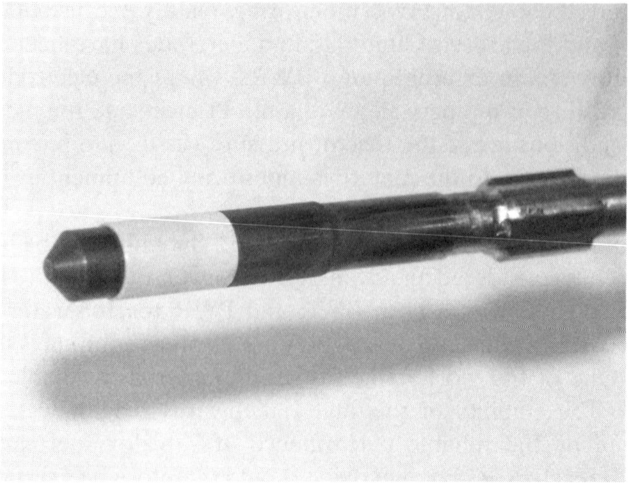

Fig. 2.10 Photograph of the Rh/Pt electrode.

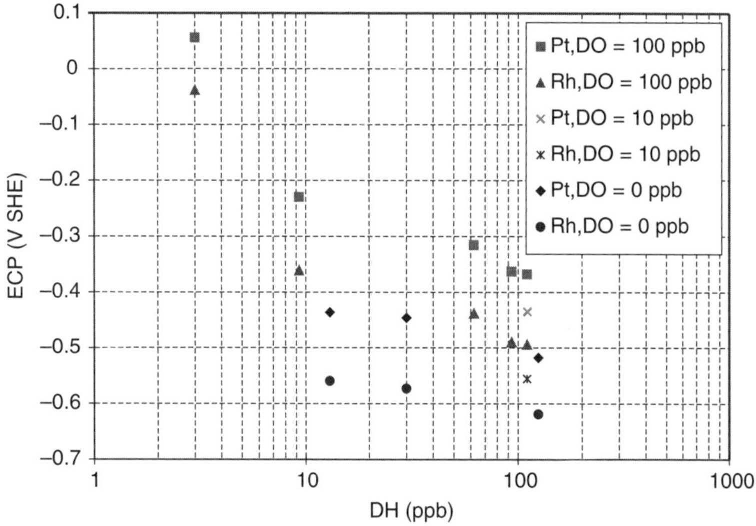

Fig. 2.11 Measured electrochemical potentials (ECP) of Rh and Pt at different combinations of H_2 and O_2. DO, dissolved oxygen concentration; DH, dissolved hydrogen concentration.

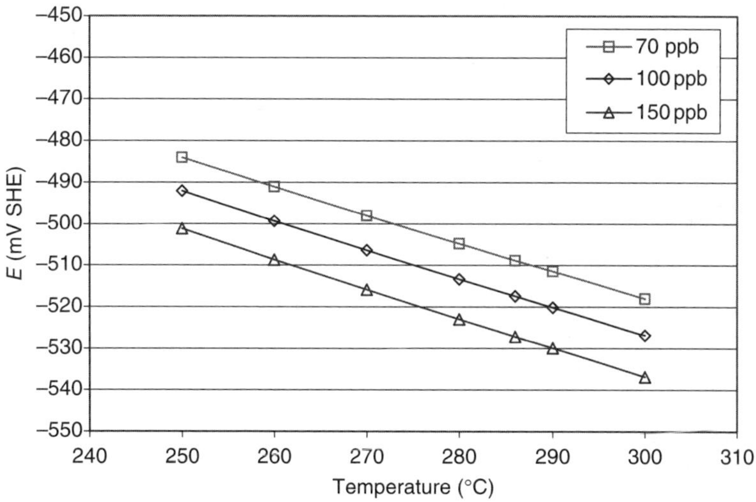

Fig. 2.12 Influence of temperature and hydrogen concentration on the potential of a Pt hydrogen electrode in neutral water.

The use of this electrode for corrosion potential measurements depends on the fact that the Pt potential is less sensitive to oxidizing species than the construction materials, i.e., the exchange current density of the hydrogen equilibrium reaction is much higher on Pt than on stainless steel.

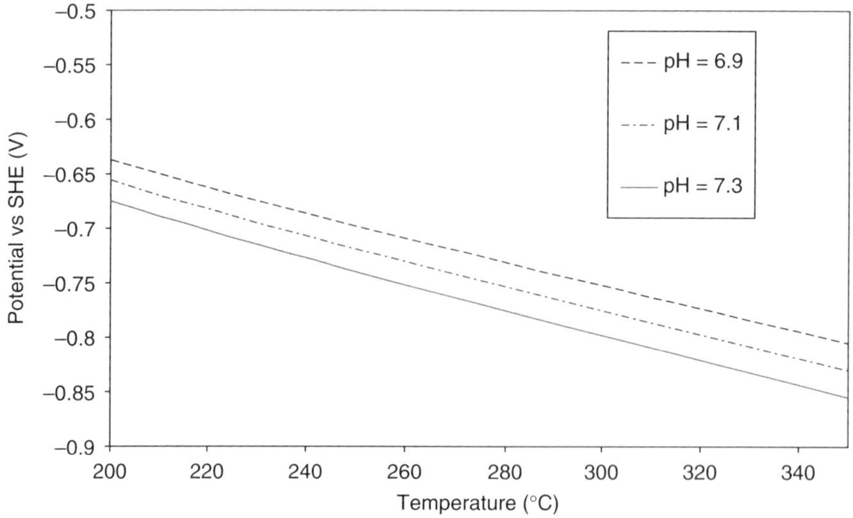

Fig. 2.13 Platinum (= Rh) potentials under pressurized water reactor conditions (2000 ppm hydrogen) for different pH values.

During weakly oxidizing conditions, the elevated corrosion potentials of the materials of construction can be measured at the almost unaltered potential of the Pt hydrogen electrode. This provides a 'window' for the practical application of the Pt hydrogen electrode and is the basis for its use as a tool to monitor HWC in BWRs.

2.3.4 Palladium electrode

The Pd-hydride reference electrode is a quasi-reference electrode where the potential only depends on the pH of the solution. The Pd electrode is polarized cathodically, thereby producing a sufficient amount of hydrogen, so that the electrode reacts like a hydrogen electrode. An in-core high temperature reference electrode based on this principle has been developed elsewhere [17, 18]. The Pd-hydride reference electrode has already been tested both under oxidizing conditions and under neutron irradiation with good results. Therefore in the round robin test of this research, only a simple version of the Pd-hydride electrode will be tested, to allow comparison with the other electrodes.

The simplified design consists of two wire electrodes: one of Pd and one of Pt. Between these two wire electrodes a constant current is applied that maintains a constant hydrogen concentration at the Pd electrode which then determines the electrochemical potential. Figure 2.14 shows a schematic of the test set-up.

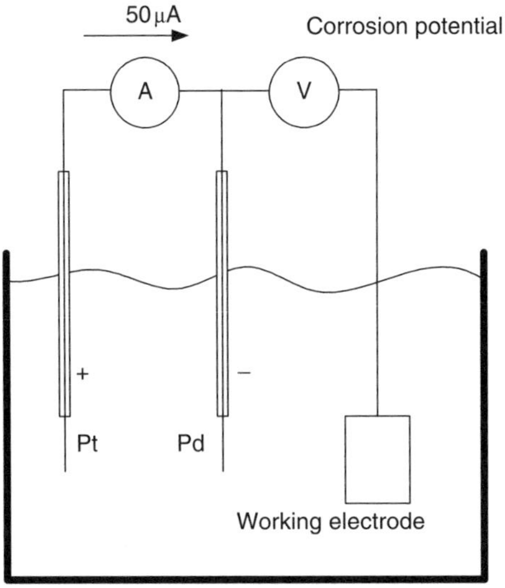

Fig. 2.14 Test set-up for measuring corrosion potential with the Pd-hydride reference electrode.

A constant current of 50 μA is applied between a Pt (anode) and a Pd (cathode) electrode. Hydrogen is produced at the Pd electrode by electrolysis of the water, which then determines the electrode potential together with the pH of the test solution. The electrode potential can be calculated with the Nernst equation for the hydrogen equilibrium reaction [16]. For example: the composition of the PWR primary water used during the round robin results in a pH of 6.9 at 300°C and contains 2 ppm hydrogen. At 300°C this results in an electrode potential of –752 mV SHE.

2.4 Round robin test

Five different laboratories have tested the four high temperature reference electrodes. Four laboratories have tested three electrodes and one laboratory has tested four electrodes. That means a total of 16 tests, i.e., each electrode was tested four times. Table 2.3 shows the distribution of the reference electrodes over the five laboratories. The developer of a high temperature reference electrode constructed four electrodes which were then distributed to the testers according to Table 2.3.

All electrodes were tested according to a previously defined test standard [5]. All the tests were performed under PWR conditions (2 ppm Li and 1200 ppm B and 2 ppm hydrogen) at three different temperatures (300, 325 and 350°C. During the test at 300°C), the water chemistry was changed from reducing conditions to oxidizing conditions and back. The objective was to

Table 2.3 Test campaign for the four different high temperature reference electrodes

Laboratory no.	Laboratory name	Type of reference electrode			
1	AEKI	Ag/AgCl	Rh	Pd	
2	CEA		Rh	Pd	YSZ
3	CIEMAT	Ag/AgCl		Pd	YSZ
4	NRI Rez	Ag/AgCl	Rh	Pd	YSZ
5	Framatone	Ag/AgCl	Rh		YSZ

Table 2.4 Potentials to be measured between the different reference electrodes

No.	Electrode I	Electrode II
1	YSZ electrode	Pt wire electrode
2	External Ag/AgCl electrode	Pt wire electrode
3	Rh electrode	Pt wire electrode
4	Pd electrode	Pt wire electrode

observe the individual responses of the reference electrodes to a change in the environment. As a first indication for the lifetime of the electrodes, a long-term test was planned at the test temperature of 300°C over a period of 3 months. The potentials were measured against a Pt electrode. At 300°C this Pt electrode consists of a small piece of Pt wire (2–3 cm) point-welded to a stainless steel wire. This stainless steel wire is then isolated with heat-shrinkable Teflon tubing. At 325 and 350°C another construction is used as Teflon cannot be used at temperatures above 300°C. A zirconium wire is point-welded to a small piece of Pt and then this combination is oxidized in an oven at ±500°C. The oxidation produces a nonconducting layer of zirconium oxide on the outside of the conducting zirconium wire.

The reference electrode potentials were all measured against a Pt electrode as common reference. Table 2.4 shows the measured potentials. So it is also possible to relate the different reference electrode potentials to each other.

The reference electrode potentials have been related to the SHE scale. Table 2.5 shows the theoretically calculated potentials of the high temperature reference electrodes versus SHE at 300°C for the water chemistry used in the round robin test [5].

Notice that the potentials of the Rh electrode and the Pd electrode are in principle equal to the potential of the Pt electrode. In reality there will

Table 2.5 Potentials of the high temperature reference electrodes against standard hydrogen electrode (SHE) for pressurized water reactor primary water chemistry (2 ppm Li and 1200 ppm B and 2 ppm hydrogen) at 300°C

Reference electrode	Potential vs SHE (mV)	Potential vs Pt electrode (mV)
Temperature	300°C	300°C
pH	6.87	6.87
YSZ electrode	−731	14
External Ag/AgCl electrode[a]	97	842
Rh electrode[b]	−745	≈0
Pd electrode[b]	−745	≈0

[a] The maximum solubility of AgCl at 25°C is taken as 1.26×10^{-5} mol kg^{-1} [19].
[b] Equal to Pt potential.

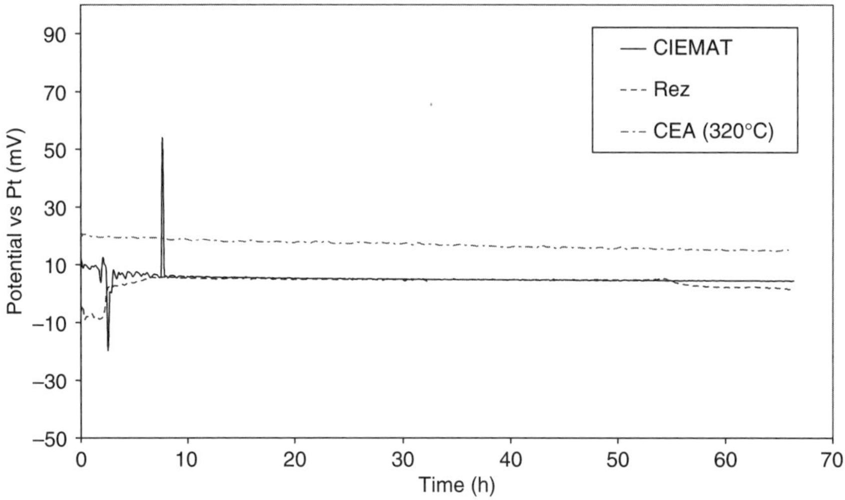

Fig. 2.15 Round robin test results of the yttrium-stabilized zirconia electrode at 300°C, pressurized water reactor primary water (2 ppm Li, 1200 ppm B and 2 ppm H_2).

always be a small difference. Figures 2.15–2.18 show the first results of the potential measurements of each high temperature reference electrode.

The potentials of the YSZ electrode and the Rh electrodes are quite stable and close to their theoretical values. Also the Pd electrode has a stable value although the differences between the three laboratories are slightly bigger. The Ag/AgCl electrode also has a potential close to the theoretical potential, but the signal of one laboratory is very noisy. This is most probably caused by small air bubbles in the capillary tube, which is

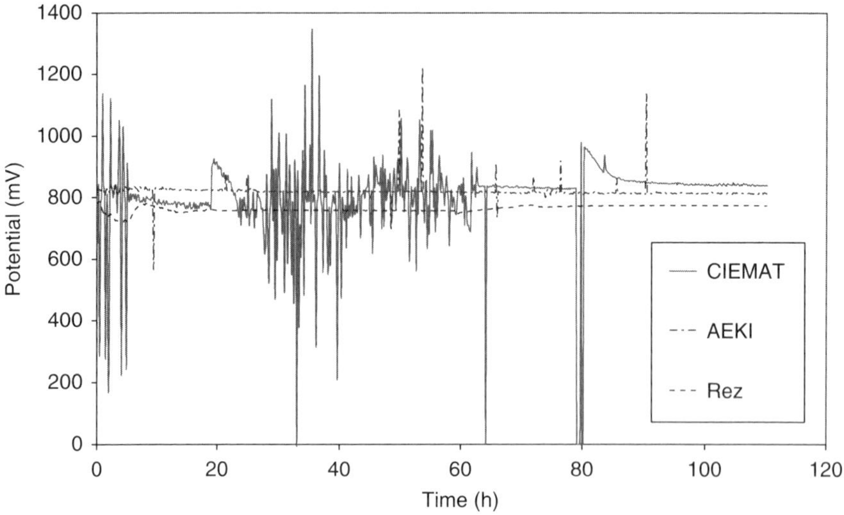

Fig. 2.16 Round robin test results of the Ag/AgCl electrode at 300°C, pressurized water reactor primary water (2 ppm Li, 1200 ppm B and 2 ppm H_2).

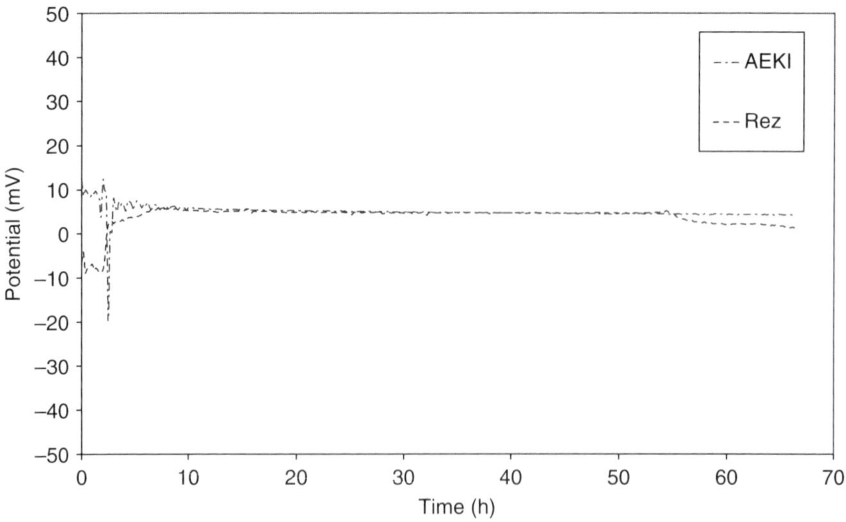

Fig. 2.17 Round robin test results of the Rh electrode at 300°C, pressurized water reactor primary water (2 ppm Li, 1200 ppm B and 2 ppm H_2).

used as a salt bridge. Then the impedance of the electrode is very high and the potential readings become noisy.

Table 2.6 summarizes the potentials of the high temperature reference electrodes collected to date. Although the potentials are sometimes

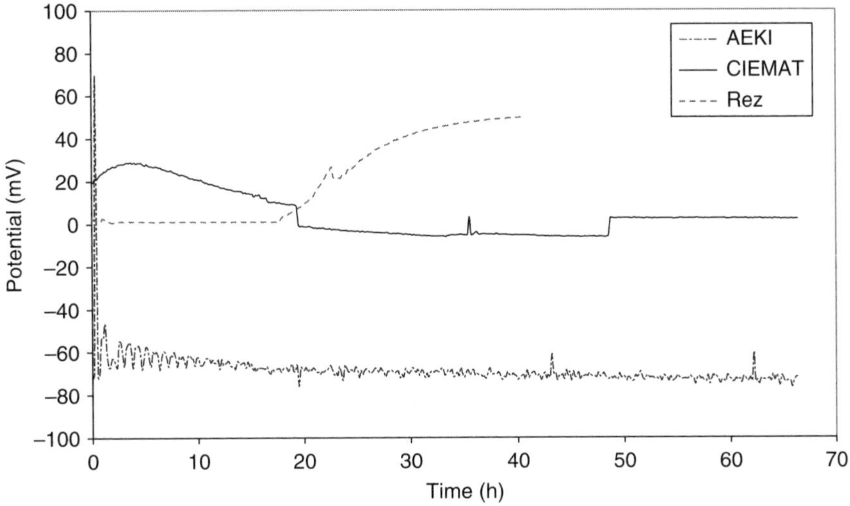

Fig. 2.18 Round robin test results of the Pd electrode at 300°C, pressurized water reactor primary water (2 ppm Li, 1200 ppm B and 2 ppm H_2).

Table 2.6 Summary of the potential measurements

Reference electrode	Measured potential vs Pt electrode (mV)	Theoretical potential (mV)
YSZ electrode	0–50	14
External Ag/AgCl electrode	750–850	842
Rh electrode	5–10	≈0
Pd electrode	−50–0	≈0

scattered, in general they represent values that are in agreement with the theoretical values.

The most important conclusions that can be drawn are as follows.

- The potentials of the high temperature reference electrodes do have proper values with respect to theoretically calculated potentials under PWR + hydrogen conditions even if the potential readings are noisy.
- In most cases, problems with the electrical isolation resistance appeared sooner or later, but they do not always influence the measured potential.
- Part of the round robin test was an oxygen trip, i.e., the water chemistry was changed from hydrogen to nitrogen to oxygen and back. The Pt

potential (oxidation reduction potential) is very sensitive to these changes, but the Ag/AgCl electrode and the YSZ electrode are not. However, the oxygen trip has so far not been very successful. The responses to changes in oxygen content of the high temperature reference electrode are not very significant. Part of the problem is related to difficulties in controlling the amount of dissolved oxygen in the test solution. If oxygen is not supplied continuously, the concentration will decrease due to oxygen consumption by the autoclave walls.

2.5 Irradiation experiments

2.5.1 Boiling water reactor conditions

The YSZ electrode and the Rh electrode have been tested in a BWR loop in the Material Test Reactor R2 (Studsvik Nuclear). Therefore a special irradiation rig has been developed with a feed-through system for the signal cables. The two electrodes were connected to a long tube with the electrodes at one end and at the other end the test rig head with the signal cable feed-throughs. Figure 2.19 shows the electrode arrangement in the test rig. The test conditions and theoretical electrochemical potentials (ECPs) are summarized in Tables 2.7 and 2.8.

ECP values were measured between the reference electrode and the stainless steel ground. Due to the low conductivity of the test solution, the working electrodes are those parts of the stainless steel tube surrounding the reference electrodes that are closest to the reference electrodes. The low conductivity of pure water means that fairly local variations of ECP level are possible. A source of such variations could be, for example, differences in corrosion rate. Higher passive current results in a lower ECP value.

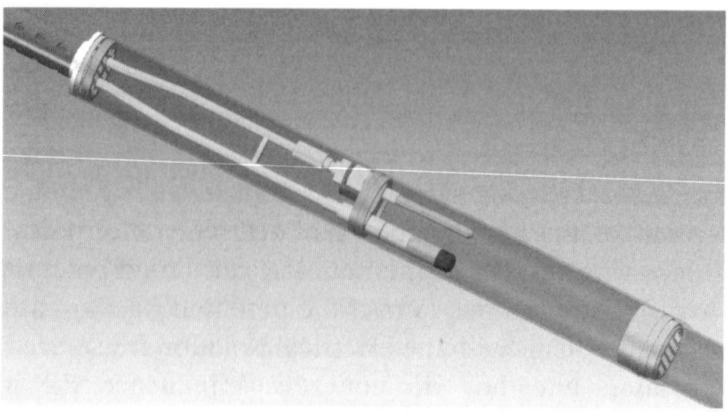

Fig. 2.19 Test rig electrode arrangement; the upper electrode is the yttrium-stabilized zirconia electrode and the lower electrode the Rh electrode.

Table 2.7 Test conditions for boiling water reactor loop

Temperature	280°C
pH at 280°C	5.7
Conductivity	±3 μS cm^{-1}
Dissolved hydrogen concentration	1 ppm
Flow rate	4 m s^{-1}

Table 2.8 Theoretical potentials under boiling water reactor conditions

Potential YSZ (Ni/NiO)	–581 mV SHE (at 280°C)
Potential Rh electrode	–573 mV SHE (at 280°C)

Temporal variations of corrosion rate, and hence of ECP, are also possible. In addition to 'real' ECP variations, electrical noise may also affect ECP measurement.

The reference electrodes have been tested during 6 reactor periods. Data from test campaign 0407 are shown in Fig. 2.20. Obviously there is no useful signal from the membrane electrode in most of this period. However, the start-up period is of interest and is shown in detail in Fig. 2.21. The loop is heated twice before the reactor is started up. Before the first heat-up, the potential of stainless steel (SS) is 300–400 mV higher than after heat-up, as measured with the Pt-Rh electrode (the membrane electrode does not operate at low temperature). This is in good agreement with the fact that oxygen enters the loop water when the loop is opened up during the shut-down. When the loop is heated up, the oxygen is consumed by corrosion and the SS ECP falls. When the temperature is lowered again, the loop water remains oxygen-free and the SS ECP stays low. From loop heat-up until reactor start-up, the potentials measured with the two reference electrodes are close together. The start-up causes the signal from the membrane electrode to increase, and then to deteriorate. Just as during period 5, the potential measured with the Pt-Rh electrode remains at around 230 mV.

Figure 2.22 shows data from the last test campaign (period 0411). The reactor power level was very stable during period 11. Despite this, the measured potentials were not stable. In this period, however, there is a strong correlation between the two measured potentials. For example, in the 5-day period from day no. 332 to day no. 337, the coefficient of correlation of the two signals is about 0.91. We assume therefore that there must be a common cause for the fluctuations of the measured potentials. For the last 6 days of the test, the potential measured with the membrane electrode is stable. The level is about 230 mV. This is the same level where the Pt-Rh electrode was stable in some of the previous test periods.

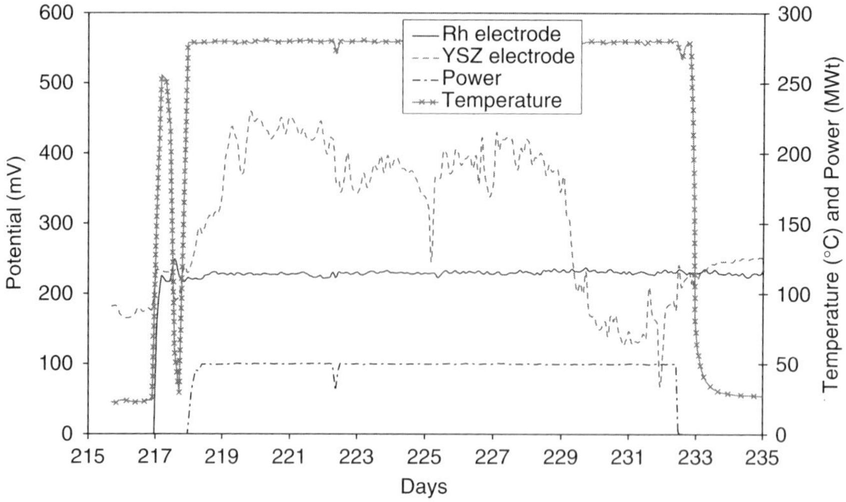

Fig. 2.20 Potentials of the Rh electrode and yttrium-stabilized zirconium (YSZ) electrode in the boiling water reactor (BWR) loop in the R2 reactor (BWR primary water, hydrogenated, 280°C) – test campaign 0407.

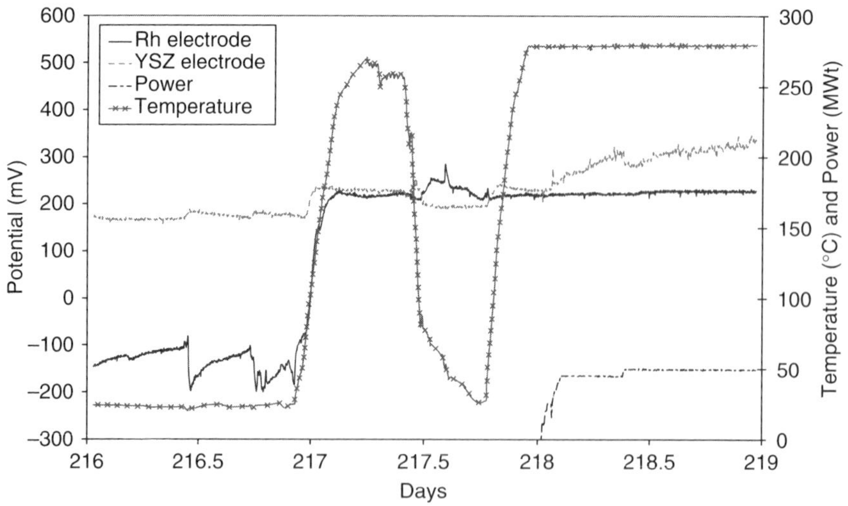

Fig. 2.21 Potentials of the Rh electrode and yttrium-stabilized zirconia (YSZ) electrode in the boiling water reactor (BWR) loop in the R2 reactor (BWR primary water, hydrogenated, 280°C) – detail of test campaign 0407.

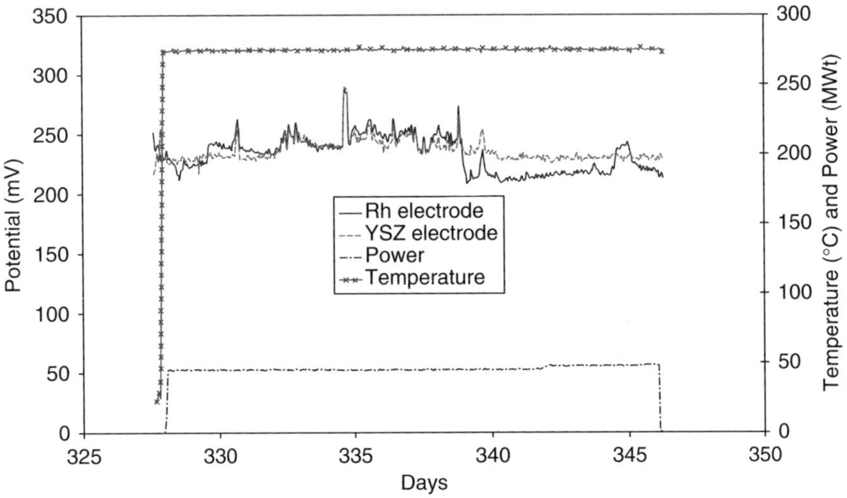

Fig. 2.22 Potentials of the Rh electrode and yttrium-stabilized zirconia (YSZ) electrode in the boiling water reactor (BWR) loop in the R2 reactor (BWR primary water, hydrogenated, 280°C) – test campaign 0411.

To summarize:

- The measured potentials are periodically stable, periodically unstable. When stable, the measured potential level is about 230 mV. This is more than 200 mV above the expected level.
- The measured potential signals are characterized by several large, sudden shifts. SS ECP is known to vary slowly.
- The potential measured with the membrane electrode is especially noisy with many transients. Some of the transients, but not all of them, seem to be initiated by variations in the reactor power level. In some cases, there is a linear correlation between the reactor power level and the measured potential during a transient.
- Normally, the signal of the YSZ electrode is uncorrelated to that of the Rh electrode (except when both are stable at around 230 mV). However, during periods 9 and 11, both channels are noisy and the signals correlated.

2.5.2 Pressurized water reactor conditions

The YSZ electrode and the Rh electrode are being tested in a PWR loop in the BR2 material test reactor. Therefore a special irradiation rig has been developed with a special feed-through system for the signal cables [20]. Three YSZ electrodes and two Rh electodes have been tested. They are

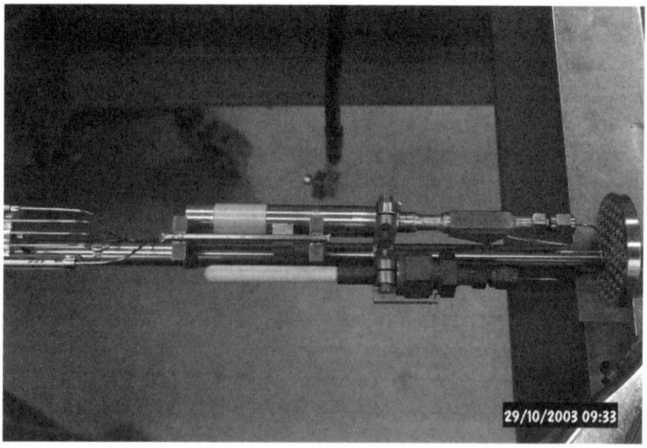

Fig. 2.23 Detail of the irradiation rig with yttrium-stabilized zirconia and Rh electrodes.

positioned influx (2 × YSZ electrode + 1 × Rh electrode) and outflux (1 × YSZ electrode + 1 × Rh electrode), to be able to investigate the influence of radiation on the electrode potentials. Figure 2.23 shows a picture of the YSZ and Rh electrode mounted in the irradiation rig. The water chemistry is slightly different from the round robin test as requested for the CALLISTO loop: PWR primary water with 2 ppm Li, 400 ppm B and 2 ppm H_2. Table 2.9 gives an overview of the test conditions.

The electrode potentials are measured against the frame of the rig and a high temperature Pt electrode. This Pt electrode consisted of a Pt cylinder connected to a mineral insulated signal cable. The feed through is a ceramic (aluminium oxide) connector with brazed metal covers at both ends. The aluminium oxide ceramic connector is surrounded with a zirconium oxide tube (similar construction as in Fig. 2.3). This construction can be used for at least 1 month (based on laboratory experiments) before the isolation resistance has decreased to an unacceptably low level due to dissolution of the ceramic isolation.

Figure 2.24 shows the potentials of the YSZ electrodes during one reactor cycle. The potentials are all close to zero. The influx electrodes tend to change with time. The electrodes respond to variations in hydrogen. Figure 2.25 shows the potentials of the Rh electrodes. With the start of the reactor they increased a little, due to the temperature increase (caused by γ-heating). The potentials are again close to their theoretical values.

Figure 2.26 shows the potential differences between the reference electrodes positioned influx and outflux. These differences are small and so we

Table 2.9 Characteristics of the pressurized water reactor loop CALLISTO

	Thermal flux (10^{14} N cm^{-2} s)	Epithermal flux (10^{14} N cm^{-2} s)	Fast flux (>100 keV) (10^{14} N cm^{-2} s)	Fast flux (>1 MeV) (10^{14} N cm^{-2} s)	Gamma heating (WG_{AL}^{-1})
Nuclear data: (at 56 MW BR2 reactor power and at mid-reactor plane)					
IPS 1 and IPS 3	0.3 . . . 1.8	0.03 . . . 0.09	0.3 . . . 0.8	0.09 . . . 0.33	1.0 . . . 3.0
IPS 2	0.6 . . . 5.0	0.25 . . . 0.40	2.5 . . . 3.5	0.5 . . . 1.2	5.0 . . . 8.0
Thermohydraulic data					
Temperature	80°C ≤ T ≤ 300°C (inlet of the IPS)				
Pressure	1.0 MPa ≤ P ≤ 15.7 MPa				
Flow rate	1.8 kg s^{-1} ≤ G ≤ 2.5 kg s^{-1}				
Chemical data					
Boron (H_3BO_3)	±400 ppm				
Lithium (LiOH)	1.8 ppm ≤ [Li] ≤ 2.2 ppm				
pH	7.00 ≤ $pH_{25°C}$ ≤ 7.08 or 7.26 ≤ $pH_{300°C}$ ≤ 7.34				
Dissolved hydrogen	25 ccSTP kg^{-1} ≤ [H_2] ≤ 35 ccSTP kg^{-1}				

can conclude that the direct effect of radiation on the electrode potential is negligible. Only the potential difference between the Pt electrodes shows a small drift over time. As the Pt electrodes are the most sensitive to changes in water chemistry, this still can be related to formation of electrochemically active radiolysis products.

Figure 2.27 shows photographs of the LIRES rig after 1 year of testing in the CALLISTO PWR loop of the BR2 material test reactor. Both the Rh electrodes and the YSZ electrodes are mechanically intact. The electrodes are covered with a layer most probably composed of corrosion products from other parts of the loop and the rig. Although the YSZ electrode looks undamaged, the impedance is low probably due to this layer. This might have an influence on the potential measurements, although the potential readings are still within the expected range even at the end of the test period.

To summarize:

- The measured potentials were stable during all test campaigns, besides test campaign 2, where the test temperature was 80°C.

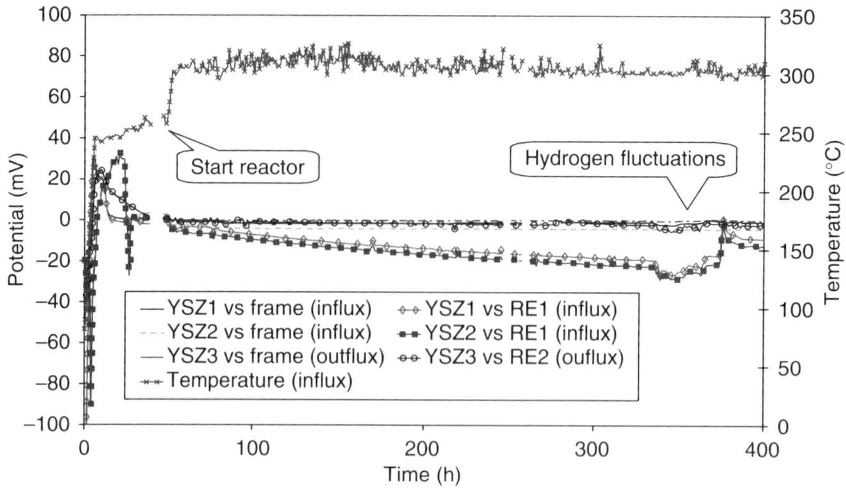

Fig. 2.24 Potentials of the yttrium-stabilized zirconia (YSZ) electrode in the pressurized water reactor (PWR) loop in the reactor, PWR primary water (2 ppm Li, 400 ppm B and 2 ppm H_2).

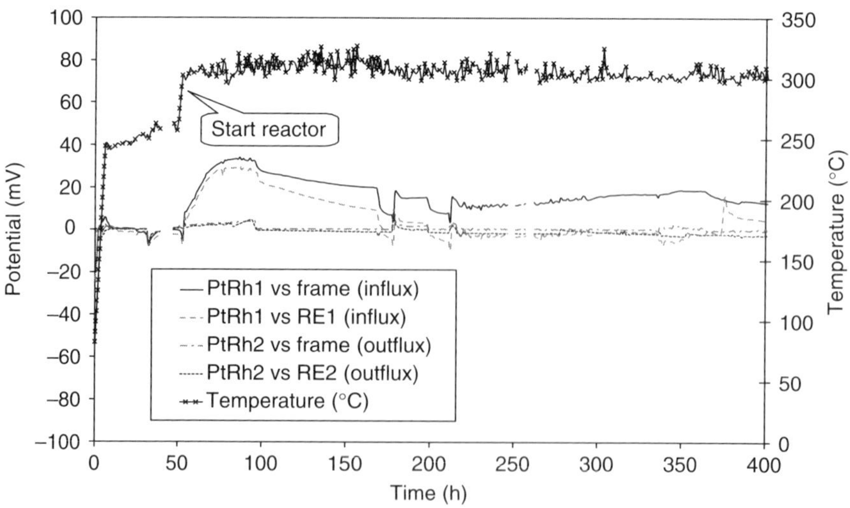

Fig. 2.25 Potentials of the Rh electrode in the pressurized water reactor (PWR) loop in the reactor, PWR primary water (2 ppm Li, 400 ppm B and 2 ppm H_2).

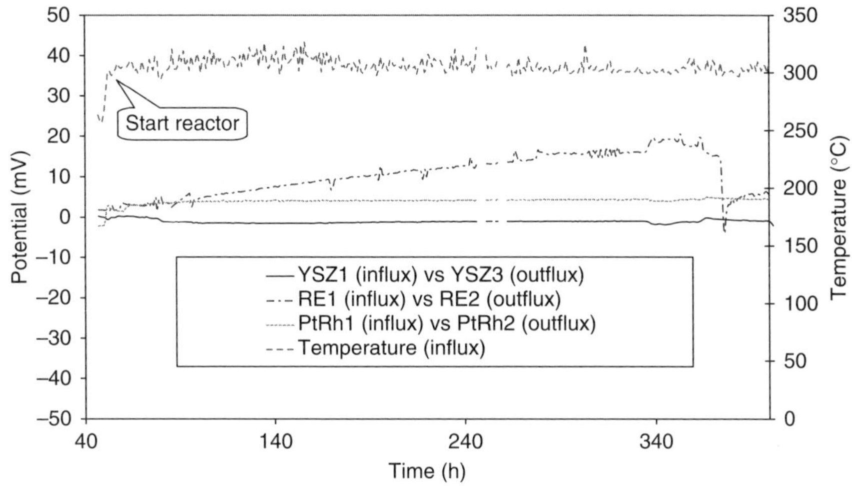

Fig. 2.26 Potential differences between the influx and outflux yttrium-stabilized zirconia (YSZ) electrodes, Rh electrodes and the Pt electrodes, pressurized water reactor primary water (2 ppm Li, 400 ppm B and 2 ppm H_2).

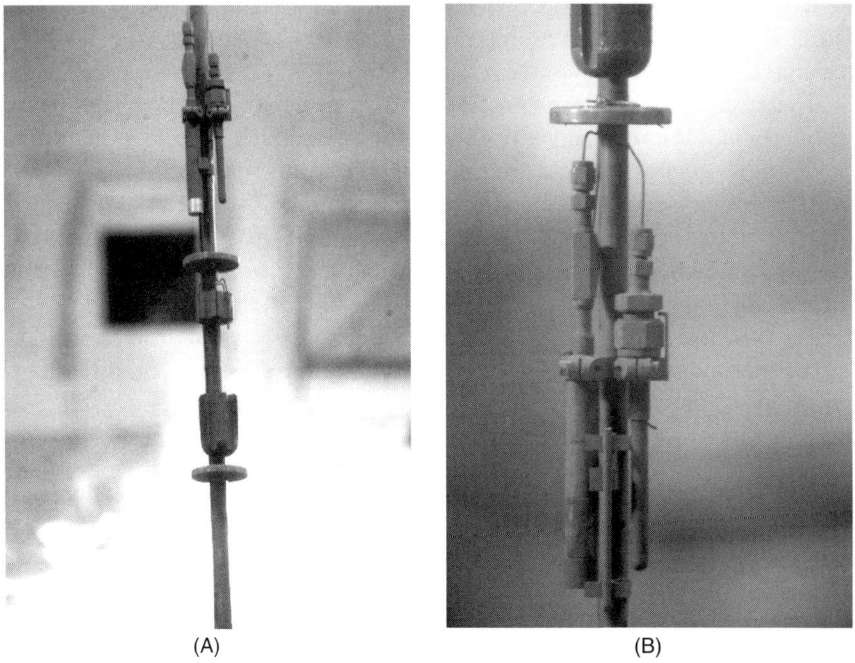

Fig. 2.27 The yttrium-stabilized zirconia electrode and the Rh electrode after 1 year of testing in the BR2 reactor. (A) Outflux position; (B) influx position.

- The potentials were in agreement with the theoretical values. However, these values were close to zero, a result which can also be obtained in the case of short circuit of the electrodes.
- The YSZ electrode followed a change in water chemistry during the first test campaign, proving proper working of the reference electrode.
- Both the YSZ electrode and the Rh electrode were mechanically intact after 1 year of testing.
- The electrochemical impedance of the YSZ electrode was very low after 1 year of testing. This could have corrupted a reliable measurement of the corrosion potential.
- One of the Rh electrodes was leaking. However, the influence on the potential measurement was small. This small effect could be diminished by draining the electrode.

2.6 Conclusions

Four potential in-core high temperature reference electrodes have been developed for use in LWRs: a YSZ electrode, an external Ag/AgCl electrode, a Rh electrode and a Pd electrode (only for the round robin). The reference electrodes are constructed from irradiation resistant materials and can be used up to 350°C, which allows them to be used inside a reactor core. The potentials of these high temperature reference electrodes can be related to the SHE scale.

The results, at laboratory scale, have shown that the electrodes work properly and are mechanically and chemically stable at the representative LWR water chemistries and operating temperatures and pressures. The four reference electrodes have been tested in a round robin test by most of the five laboratories. The results collected show that the measured potentials are in agreement with each other and with theoretically calculated potentials.

The YSZ electrode and the Rh electrode have been tested in an irradiation experiment under BWR and PWR conditions. The potential readings are in agreement with theoretical values and did not show any degradation despite the presence of the high radiation field and the high testing temperature. Visual inspection showed that both the YSZ and the Rh electrodes are mechanically intact. However, the impedance of the YSZ electrode was small, which may hamper the correct measurement of electrochemical potentials.

2.7 Acknowledgment

This work was performed with financial support from the European Commission, EURATOM FP5, contract no. FIKS-CT-2000-00012.

2.8 References

1. P. Scott, 'A Review of Irradiated Assisted Stress Corrosion Cracking', *J. Nucl. Mat.* **211**, 101 (1994).
2. P. Scott, 'Stress Corrosion Cracking in Pressurized Water Reactors – Interpretation, Modeling and Remedies', *Corrosion* **56**, 771 (2000).
3. M. E. Indig, J. L. Nelson, 'Electrochemical Measurements and Modeling Predictions in Boiling Water Reactors Under Various Operating Conditions', *Corrosion* **47**, 202 (1991).
4. J. H. Zheng, W. F. Bogaerts, 'Reference Electrodes for Electrochemistry and Corrosion Study in High Temperature Aqueous Environments', EC report, LIRES project, University of Leuven, Belgium (2001).
5. R.-W. Bosch *et al.* LIRES: A European Sponsored Research Project to Develop Light Water Reactor Reference Electrodes, *Int. Conf. Environmental Degradation of Materials in Nuclear Power Systems – Water Reactor*, 10–14 August, 2003, Stevenson, WA.
6. R.-W. Bosch *et al.* LIRES: Actual Status of the LIRES-project: Development of Light Water Reactor Reference Electrodes, EUROCORR, 28 September–2 October, 2003, Budapest, Hungary.
7. L. W. Niedrach, 'Oxygen Ion-Conducting Ceramics: A New Application in High-Temperature High-Pressure pH Sensors', *Science* **207**, 1200 (1980).
8. L.W. Niedrach, 'A New Membrane-Type pH Sensor for Use in High Temperature High Pressure Water', *J. Electrochem. Soc.* **127**, 1445 (1980).
9. D. D. Macdonald, S. Hettiarachchi, H. Song, K. Makela, R. Emerson, M. Ben-Haim, 'Measurement of pH in Subcritical and Supercritical Aqueous Systems', *J. Solution Chem.* **21** 849 (1992).
10. K. Mäkelä, Unpublished data VTT (2002).
11. M. E. Indig, 'Technology Transfer: Aqueous Electrochemical Measurements Room Temperature to 290°C', *Corrosion* **46**, 680 (1990).
12. M. J. Danielson, 'The Construction and Thermodynamic Performance of an Ag-AgCl Reference Electrode for Use in High Temperature Aqueous Environments Containing H_2 and H_2S', *Corrosion* **35**, 200 (1979).
13. D. D. Macdonald, 'Reference Electrodes for High Temperature Aqueous Systems – A Review and Assessment', *Corrosion* **34**, 75 (1978).
14. D. D. Macdonald, A. C. Scott and P. Wentrcek, *J. Electrochem. Soc.* **126**(5), 908 (1979).
15. R.-W. Bosch, W. F. Bogaerts, J. H. Zheng, 'Simple and Robust External Reference Electrodes for High-Temperature Electrochemical Measurements', *Corrosion* **59**, 162 (2003).
16. *Corrosion Potential (ECP) Measurement Sourcebook*, EPRI NP-7142, Electric Power Reserch Institute, CA (January 1991).
17. P. J. Bennet, E. Hauso, N. W. Hogberg, T. M. Karlsen, M. A. McGrath, 'In-core Materials Testing Under LWR Conditions in the Halden Reactor', *Proc. Int. Conf. Water Chemistry in Nuclear Reactor Systems*, 21–25 April, 2002, Avignon, France.
18. B. Beverskog, L. Lie, N. W. Hogberg, K. Makela, 'Verification of the Miniturised In-core Pd Reference Electrode Operation in Out-of-core and In-core Location in Halden', *Proc. Int. Conf. Water Chemistry in Nuclear Reactor Systems* 8 BNES, 22–26 October, 2000, p. 281 (2000).

19. P. W. Atkins, *Physical Chemistry*, Oxford University Press, Oxford (1990), p. 955.
20. M. Wéber, R.-W. Bosch, L. Vermeeren, 'Design and Development of a Multipurpose PWR Rig for In-Core Instrumented Testing', *Proc. Int. Conf. Water Chemistry of Nuclear Systems*, 11–14 October, 2004, San Francisco (2004).

3

In-plant corrosion potential measurements in light water reactor environments

ANDERS MOLANDER, Studsvik Nuclear AB, Sweden

3.1 Introduction

Electrochemical monitoring is the only method available to perform actual in-situ measurements in high temperature water. All other techniques rely on sampling and cooling of a limited flow of water before some kind of subsequent measurement is performed, but electrochemical measurements can be performed directly inside the system part itself. For this reason, measuring equipment has been developed for boiling water reactor (BWR), pressurized water reactor (PWR) primary and secondary systems as well as other types of reactors.

In this chapter, equipment used in the plants is described and some examples of the results are given. The increased understanding obtained over the years concerning the electrochemical reactions determining the corrosion potential in different environments is described. This increased understanding has been used to develop models to predict corrosion potentials in the systems under study.

3.2 Measurements in boiling water reactors

Electrochemical corrosion potential (ECP) measurements were first developed for BWRs with the prime goal to follow the corrosion potential during hydrogen water chemistry (HWC). Early measurements were performed using equipment installed in systems remote from the actual reactor component [1]. It was gradually understood that too low ECPs were measured at such locations due to hydrogen peroxide decomposition, oxygen consumption and mass transfer effects [2]. With time, better techniques were developed. ECP monitoring developed to a routine power plant technology necessary for HWC operation and essential for obtaining inspection relief [3].

The improved technique used was primarily the in-pipe reference electrode as illustrated in Fig. 3.1 [4]. This is usually a Pt electrode installed

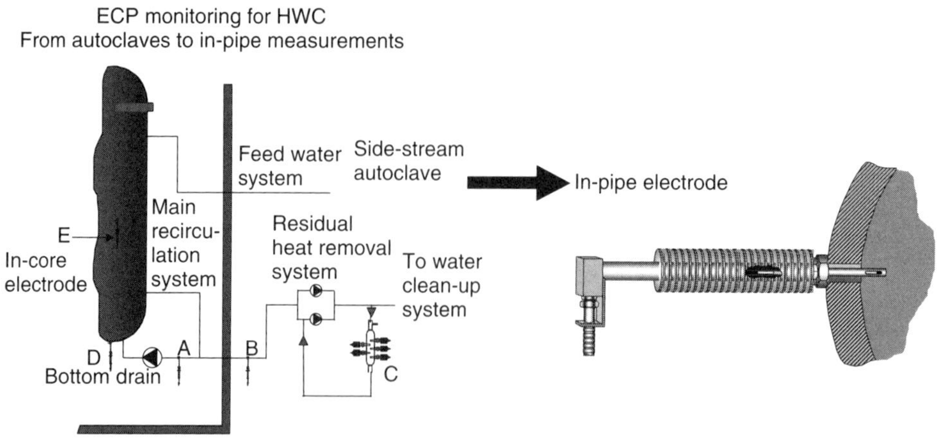

Fig. 3.1 In boiling water reactors the trend has been to find monitoring locations for electrochemical corrosion potential (ECP) measurements without sampling. The effects of the location of the measurements were noted early and explained by H_2O_2 decomposition, mass transfer effects and O_2 consumption, especially with autoclaves connected with low diameter sampling lines with low flow rate. HWC, hydrogen water chemistry.

directly into the reactor piping. The dimension of these pipes is from 250 to 600 mm. The flow rate in the pipes is 2–4 m s^{-1}. By such an electrode, the ECP of the actual reactor component can be measured and sampling is not needed. Autoclaves are not needed either.

In Fig. 3.2, some possible monitoring points in a BWR are shown. In an external pump reactor, the main recirculation system is usually the prime area to protect by HWC and the amount of hydrogen is adjusted according to the potential measured in the primary loop recirculation system less than a second transport time from the vessel (position A in Fig. 3.2). As back-up and complement, measurements can also be performed in the reactor water clean-up system which connects to the main recirculation system. Such a position (position B in Fig. 3.2) is 10–20 s away, but the pipe diameter is still 200 mm or more. Side-stream autoclaves (position C in Fig. 3.2) are no longer used for HWC surveillance.

In internal pump reactors, the bottom of the vessel is often the prime area to protect by HWC. Here, monitoring sometimes is performed in the bottom drain system (position D in Fig. 3.2) but also directly inside the vessel. This is a potential use of the LIRES electrodes as well as for in-core measurements (position E in Fig. 3.2).

Figure 3.3 shows the design of an in-pipe Pt electrode in more detail. The first electrodes used were installed in existing pressure-monitoring

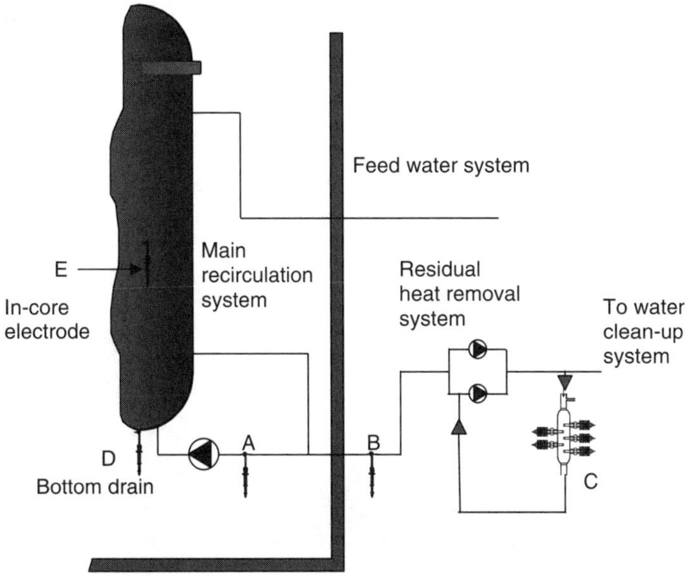

Fig. 3.2 Illustration of common monitoring points in a boiling water reactor.

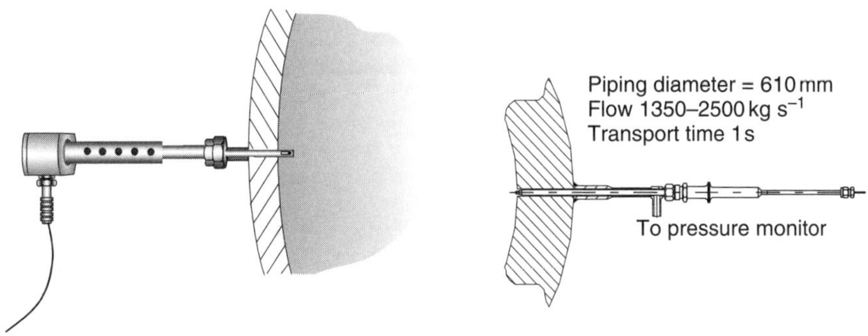

Fig. 3.3 Illustration of an in-pipe electrode for boiling water reactors.

connections. By welding a T-fitting to the connector, the electrode could be installed and also the pressure monitoring could be kept. Later, electrodes have also been installed using new connections. Then, a similar pipe to the pressure connection has been welded to the existing piping and a hole through the pipe has been drilled for the electrode to be installed into. This kind of electrode is used for 2 or 3 years depending on the location where the electrode is installed. It has a number of seals for safety. A leakage can cause a cold shut-down of the plant.

The platinum electrode described above is possible to use as a pseudo-reference electrode when a sufficient amount of hydrogen is present in the reactor water. For very low hydrogen additions or for normal water chemistry, a silver chloride electrode has also been developed for in-pipe use. The electrode is illustrated in Fig. 3.4. This electrode has a shorter lifetime compared to the platinum electrode described above.

When in-pipe measurements are impossible or unsuitable for different reasons, equipment in by-pass systems is still sometimes used. An example of equipment that can be used for such measurements is shown in Fig. 3.5.

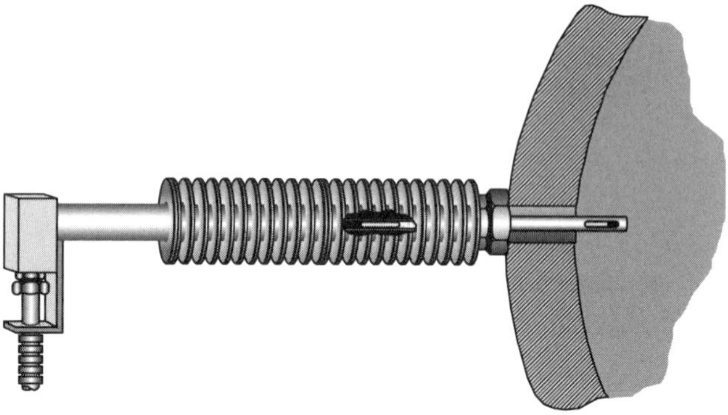

Fig. 3.4 Silver chloride electrode for in-pipe installation.

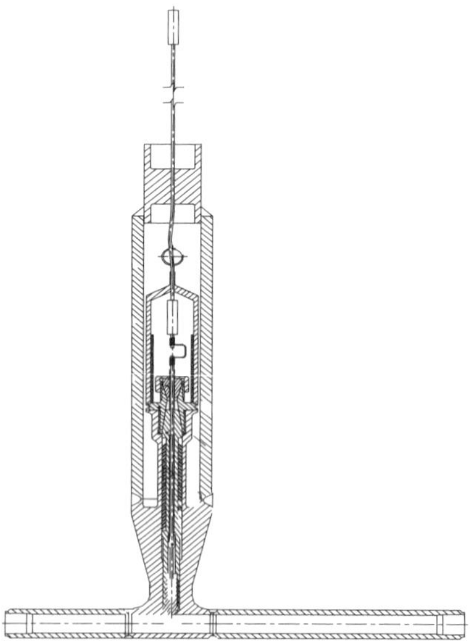

Fig. 3.5 Silver chloride electrode for high flow measurements in by-pass or sampling system areas not accessible during operation.

The figure shows a silver chloride electrode welded to a piece of piping to be connected to a by-pass or sampling system, preferably with a high flow rate. Such equipment includes double pressure boundaries allowing installation in an area that is not accessible during operation.

Sometimes even the equipment described above for use in by-pass systems or sampling systems may not be adequate and autoclaves are still used for some applications. Such equipment is used only in areas accessible during operation.

Equipment in by-pass and sampling systems has been used in power plants since 1978. In-pipe electrodes were installed for the first time in a power plant in 1988. Studsvik has been involved in BWR measurements in more than 15 BWRs. Similar equipment has also been used in RBMK reactors and in material-testing reactors.

From the in-plant measurements and also from laboratory measurements, an improved ECP model for BWR environments was developed [5]. The model provides a mechanistic background to the measured ECPs with some new features. The user can easily change input values such as the flow rate, pipe geometry and chemistry conditions and immediately see the effect on the ECP. Thus, it is called a virtual ECP laboratory. The results of the model calculations are given as Evans diagrams as shown in Fig. 3.6 for a fairly high H_2O_2 concentration. The right part of the solid line shows the hydrogen peroxide oxidation to oxygen and the left part of the dashed line shows the hydrogen peroxide reduction to water. The calculated ECP is shown as the dashed line at the intersection of thc two curves. The dashed and dotted line is the passive current of stainless steel. This Evans diagram explains the plateau regime of the ECP in water with hydrogen peroxide. A concentration change will only cause a vertical shift of the two curves, leaving the ECP unaffected.

With the Studsvik ECP model, plots of the ECP vs hydrogen peroxide concentration, for example, can be made for different conditions. In Fig. 3.7, good fit is obtained between the calculated and the measured ECPs during certain conditions. Both the plateau level and the abrupt drop to a low potential is reproduced [5].

In summary, ECP monitoring for BWRs has developed to a routine power plant technology beneficial for reactor operation and essential for obtaining inspection relief and extended inspection intervals. The lifetime of the in-pipe electrodes is 2–3 years depending on the installation location. With ECP monitoring the hydrogen addition can be optimized. This is impossible by modeling only. However, modeling of chemical and electrochemical conditions in reactor systems is important, as actual measurements cannot be performed at every point of interest. Modeling and experiment support each other and are both necessary ingredients in this type of work.

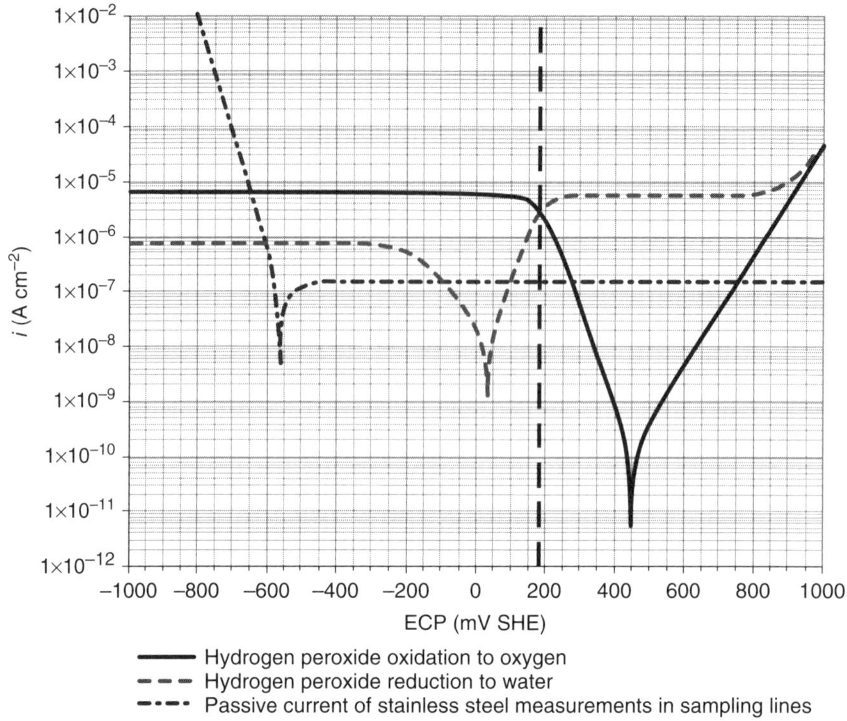

Fig. 3.6 Schematic Evans diagram for a boiling water reactor environment with 100 ppb of H_2O_2. ECP, electrochemical corrosion potential; SHE, standard hydrogen electrode. The right part of the solid line shows the hydrogen peroxide oxidation to oxygen. The left part of the dashed line shows the hydrogen peroxide reduction to water. The dashed and dotted line is the passive current of stainless steel measurements in sampling lines.

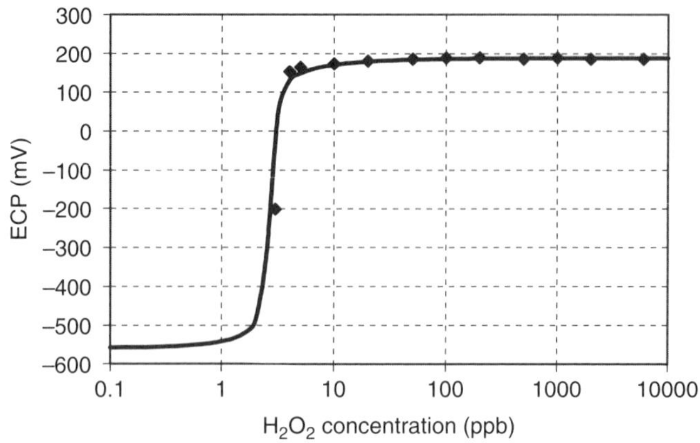

Fig. 3.7 Comparison of Studsvik experimental results (dots) with Studsvik ECP model (line) [5].

3.3 Measurements in pressurized water reactor secondary systems

Electrochemical measurements have been performed both in PWR primary [6, 7] and secondary systems [8, 9]. Here, only PWR feedwater monitoring of the secondary system is described. Such measurements were initiated later than in BWRs and the database for PWRs is smaller than the BWR database. Studsvik has been involved in secondary side measurements in six PWRs.

When these efforts were started in the beginning of the 1990s the experience from BWRs was kept in mind when designing the measurement system. In spite of that, autoclave equipment was used for practical reasons for the first measurements and three autoclaves were installed at different points in the secondary system of Ringhals 3. The sampling lines were kept as short as possible and were not longer than 1 m. A schematic design of such an autoclave and the electrodes used is shown in Fig. 3.8. Later, an in-pipe electrode for secondary side conditions was also developed. The design and the installation point are shown in Fig. 3.9. The in-pipe electrode used in Ringhals was installed close to the sampling line for the third autoclave. The installation was made in a pressure-monitoring gauge directly into the feedwater piping. The electrode comprises a Pt electrode and a silver chloride electrode.

In Fig. 3.10, an example of the results is given. During normal steady operation, there is only a very small difference between the two potentials, but when an oxidant transient occurs the in-pipe electrode reacts much more strongly than the electrode exposed in the autoclave. For the example in Fig. 3.10, the potential measured with the in-pipe electrode goes up 150 mV and the autoclave electrode shows a much weaker response.

The results of the secondary side measurements can be summarized as follows:

1. Electrochemical measurements are more sensitive to redox variations than other methods used in-plant.
2. Fast response to transients was obtained. Transients not detected by other measurements were identified.
3. In-pipe monitoring is more sensitive than measurements in autoclaves.
4. Reliable service has been demonstrated.
5. Electrochemical monitoring offers improved water chemistry surveillance and is an interesting method for routine use.

However, ECP measurements have not developed to such a routine method in PWRs in the same way as for BWRs on HWC. One reason for this is a very stable reducing chemistry in many PWRs. During such conditions without transients, the potentials tend to be low and stable.

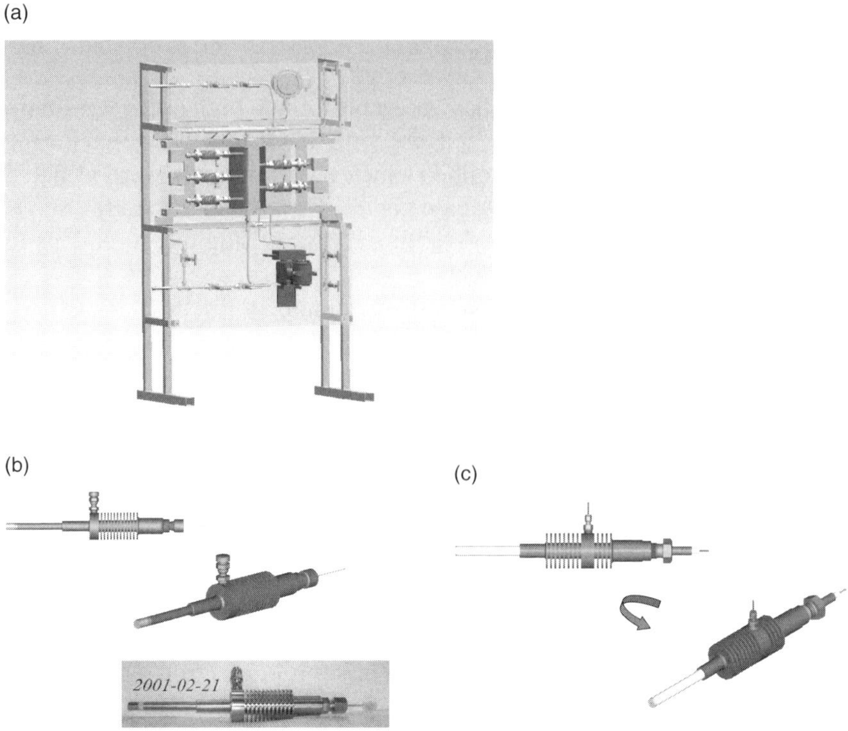

Fig. 3.8 Schematic design of autoclave (a), working electrode (b) and reference electrode (c) for in-plant measurements of redox and corrosion potentials.

Recently, but based on earlier experience, a modified type of water chemistry called oxygenated water chemistry (OWC) has been suggested [10]. The aim of OWC is to improve on current methods of limiting iron transport to the steam generators. The dominant source of iron is carbon steel used as a structural material in the secondary system. Carbon steel corrodes and releases corrosion products to the coolant due to flow-assisted corrosion (FAC), under all-volatile treatment (AVT) chemistry. OWC has successfully been adopted in BWRs and some fossil units to suppress FAC of carbon steel. Figure 3.11 illustrates the OWC concept. Oxygen is added to the feedwater line and oxygen will be gradually consumed according to the different curves shown. Curve C shows a too low addition and curve A a too high addition. Curve B would be the optimum according to this concept.

For modeling work a so-called 'OWC simulator' has been developed which in its structure is similar to the BWR ECP model. In Fig. 3.12 an example of calculations using the OWC simulator shows the decrease of

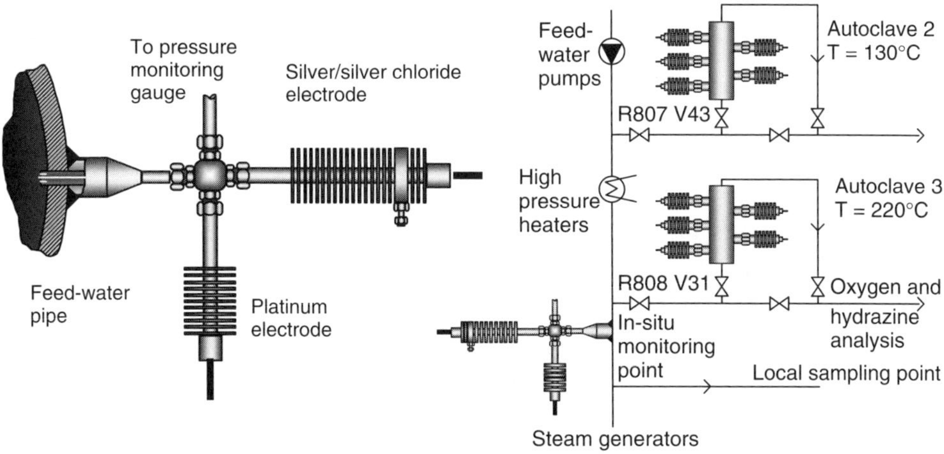

Fig. 3.9 Illustration of the in-pipe electrode developed for the secondary side of pressurized water reactors (PWRs) and the installation point in the Ringhals PWRs.

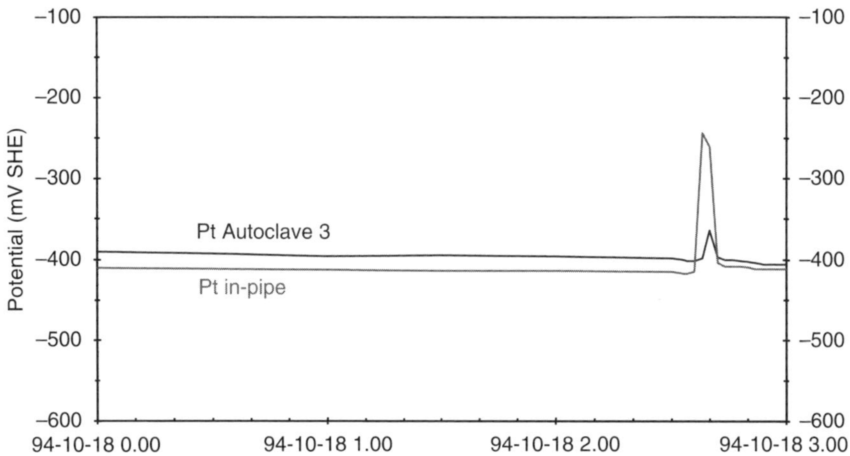

Fig. 3.10 The response to an oxygen transient as measured using the in-pipe electrode compared to the autoclave measurement [9]. SHE, standard hydrogen electrode.

the oxygen content and the corrosion potential in a 25 m long sampling line with a 10 mm diameter at 180°C with a flow rate of 20 g s^{-1}. The calculation shows that after 10 m there is no oxygen left. Due to the heterogeneous nature of the oxygen and hydrazine reaction, the oxygen decrease is very fast in a small diameter sampling line, as in the example given in Fig. 3.12. In large diameter piping the reactions occur more slowly. Thus, ECP monitoring is necessary to demonstrate OWC conditions.

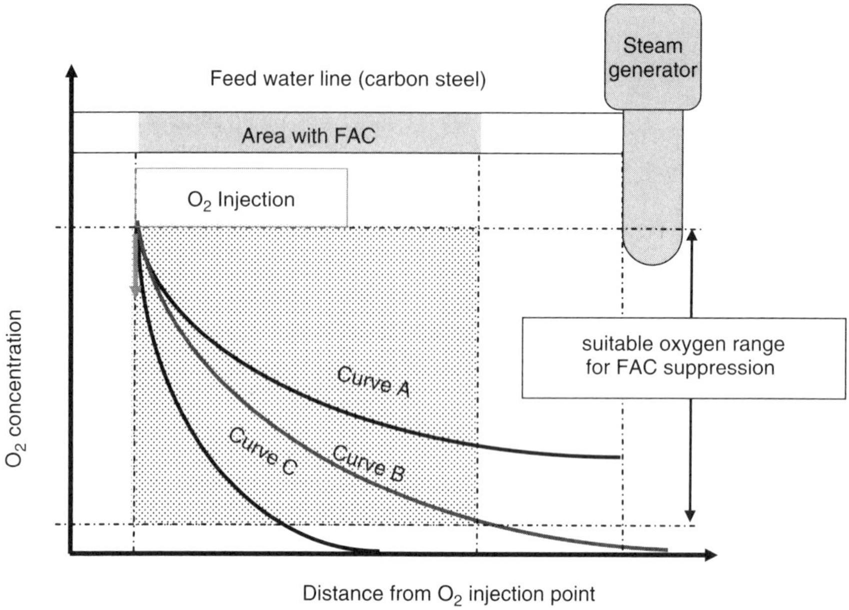

Fig. 3.11 Schematic illustration of the oxygen water chemistry concept [10]. FAC, flow-assisted corrosion.

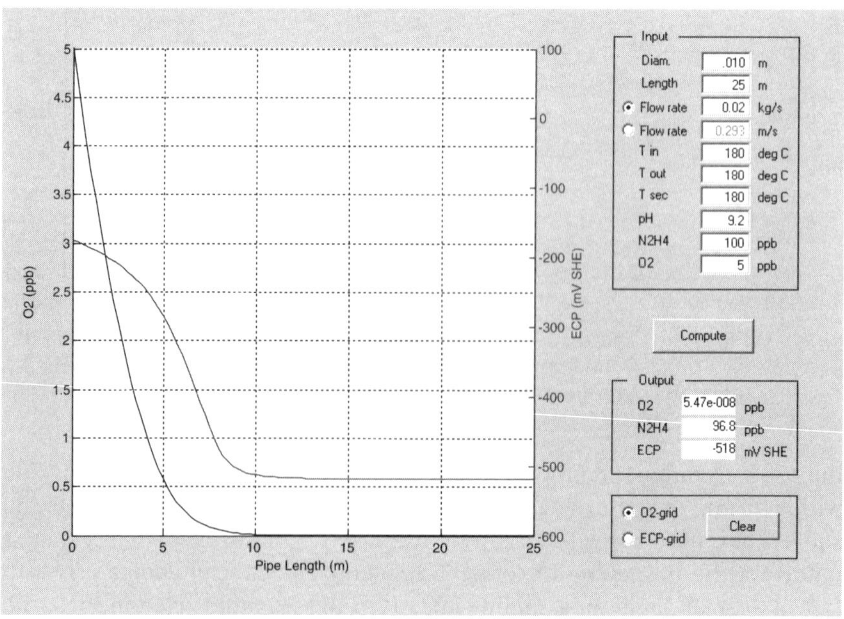

Fig. 3.12 An example of a calculation sheet from the secondary system model 'OWC (oxygen water chemistry) simulator' [10].

3.4 Conclusions

The present status of in-plant ECP measurements can be summarized as follows:

1 In BWRs electrochemical monitoring is an extremely valuable tool to monitor and control HWC.
2 In PWR primary systems, redox variations are small outside the core. For low hydrogen chemistry, electrochemical monitoring is of considerable interest.
3 In PWR secondary systems, electrochemical monitoring is a versatile tool for redox monitoring and redox mapping. Electrochemical measurements are strongly recommended for any OWC application in PWR secondary systems.

3.5 Acknowledgments

The contributions of many colleagues at Studsvik and at Swedish power plants are gratefully acknowledged, as is financial support from Swedish utilities, Swedish Nuclear Power Inspectorate and The Japan Atomic Power Company.

3.6 References

1. Rosborg, B. and Molander, A. The corrosion potential of Type 304 stainless steel in Swedish LWRs during steady operation, *Proc. Second Int. Symp. Environmental Degradation of Materials in Nuclear Power Systems – Water Reactors*. American Nuclear Society, La Grange Park, IL (1986).
2. Ullberg, M. On corrosion potential measurement in BWRs, *Proc. Fourth Int. Symp. Environmental Degradation of Materials in Nuclear Power Systems – Water Reactors*. NACE, Houston, TX (1990).
3. Molander, A. *et al.* Corrosion potential monitoring in Swedish BWRs on hydrogen water chemistry, *Proc. Ninth Int. Symp. Environmental Degradation of Materials in Nuclear Power Systems – Water Reactors*. TMS, Warrendale, PA (1999).
4. Molander, A. and Karlberg, G. Hydrogen water chemistry surveillance in a boiling water reactor, *Proc. Fourth Int. Symp. Environmental Degradation of Materials in Nuclear Power Systems – Water Reactors*. NACE, Houston, TX (1990).
5. Molander, A. and Ullberg, M. The corrosion potential of stainless steel in BWR environment – comparison of data and modeling results. *Proc. Symp. Water Chemistry and Corrosion in Nuclear Power Plants in Asia 2003*. Atomic Energy Society of Japan, Tokyo (2003).
6. Molander, A. *et al.* Corrosion potential measurements in reactor water of a PWR, *Proc. Fourth Int. Conf. Water Chemistry of Nuclear Reactor Systems,* Bournemouth. BNES, London (1986).

7. Molander, A. *et al.* Significance of corrosion potential monitoring in a PWR primary system, *Proc. Fifth Int. Conf. Water Chemistry of Nuclear Reactor Systems,* Bournemouth. BNES, London (1990).
8. Molander, A. *et al.* Electrochemical measurements in secondary system of Ringhals 3 PWR, *Proc. Sixth Int. Conf. Water Chemistry of Nuclear Reactor Systems,* Bournemouth. BNES, London (1992).
9. Molander, A. *et al.* Studies of redox conditions in feedwater of PWR secondary systems, *EUROCORR '96*. Société de Chemie Industrielle, Paris (2004).
10. Takiguchi, H., Kadoi, E. and Ullberg, M. Study on application of oxygenated water chemistry for suppression of flow assisted corrosion in secondary systems of PWRs, presented at 14th Int. Conf. Properties of Water and Steam, Kyoto, Japan 2004. Maruzen Co. Ltd, Tokyo (2005). (Also available at http://www.iapws.jp/proceedings.html.)

4
High temperature reference electrodes: a comparative analysis

MARTA NAVAS and MARISOL GARCÍA,
CIEMAT, Spain

4.1 Introduction

In making electrochemical corrosion potential (ECP) measurements at high temperature, the main difficulty is the development of a reliable reference electrode. The design of high temperature reference electrodes was very important during the 1980s, and many types were developed, mainly based on three types: noble metal electrode; Ag/AgCl electrode, internal or external; and yttrium-stabilized zirconia (YSZ) ceramic membrane electrode with different metal/metal oxide mixtures. However, great care is required to assure that valid and useful data are being obtained, because many of them still have limitations in their uses.

The requirements to obtain reliable data with a reference electrode have been extensively studied [1, 2]. In the case of high temperature reference electrodes, they can be summarized as follows:

1 Stable and constant electrode potential.
2 Chemical and thermodynamic stability.
3 Reproducible measurements.
4 Electrode potential related to standard hydrogen electrode (SHE).
5 Electrode potential independent of the environment.
6 Design requirements: system integrity and material stability.

These requirements are not only related to the thermodynamic and electrochemical behaviours, like any reference electrode; when these electrodes are going to be used at high temperature, particular emphasis must be placed on the system integrity. Several designs of high temperature reference electrodes are available for measurement under boiling water reactor (BWR) operating conditions [3]. However, the working conditions for a pressurized water reactor (PWR) reference electrode are more severe than for a BWR reference electrode. As the working temperature is higher (exceeding 300°C), different design requirements are needed.

In addition, the requirements can be different depending on the future use of the electrode. An easily fabricated electrode could be suitable for use in a laboratory, although in the case of a reference electrode for in-core electrochemical measurements it is necessary to have a robust electrode with a long life.

This chapter includes the experience obtained at CIEMAT, both in previous tests and in the round robin test of the LIRES project (Development of Light Water Reactor Reference Electrodes). First tests have been performed under simulated BWR operating conditions (290°C) with reference electrodes developed at CIEMAT. Reference electrodes of the round robin tests have been tested under PWR conditions with temperatures of 300–340°C.

4.2 Description of the work

The objective of this experimental work is to check the validity of different reference electrodes based on the experimental results. The qualification of reference electrodes has been performed by means of the measurements of electrode potential and corrosion potential with several reference electrodes. A comparison of the different types of reference electrodes simultaneously during service can indicate malfunctioning or good behaviour. The tests have been carried out under BWR and PWR operating conditions.

4.2.1 Tests under boiling water reactor operating conditions

High temperature reference electrodes

In this study three types of reference electrodes have been fabricated with some modifications and improvements with respect to the electrodes described in the literature, as follows:

1 The YSZ electrode consists of an yttrium-stabilized zirconia closed-end tube, filled with an internal junction of copper/cuprous oxide solid mixture (Fig. 4.1). It can serve as a membrane-type electrode because there is a direct relationship between the conduction of oxygen ions through the ceramic and the pH in the adjacent aqueous phase [4]. In this case, the electrode works as a membrane-type pH sensor and, when the pH of the system is known and constant, it acts as a reference electrode.
2 The Ag/AgCl internal electrode is a silver wire with electrodeposited silver chloride, enclosed in a solid Rulon chamber, filled with demineralized water (Fig. 4.2). Some silver chloride crystals are added in order to avoid changes in silver chloride solubility with temperature. As long as

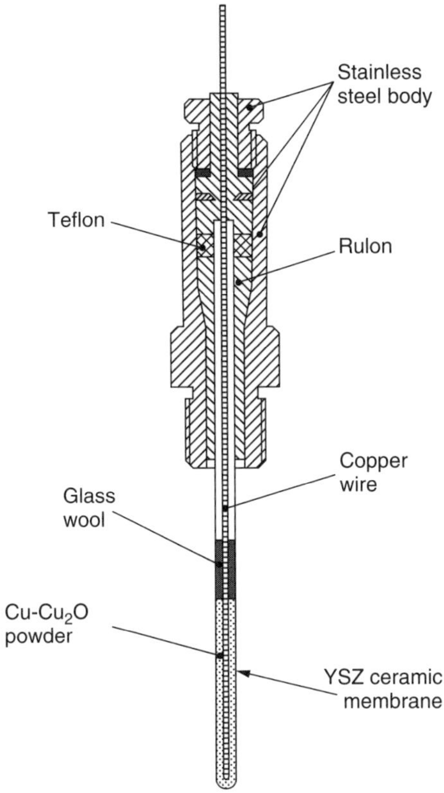

Fig. 4.1 Yttrium-stabilized zirconium (YSZ) reference electrode.

the solubility of silver chloride is known, the potential of the reference electrode is easily calculated in an SHE scale, using the Nernst equation.

3 The platinum electrode consists of a platinum wire joined to a stainless steel wire covered with heat-shrinkable Teflon. The pressure seal was made with Rulon in the same manner as the previous electrodes. Platinum wire can be used as a reference only when there is a stoichiometric excess of dissolved hydrogen concentration. This sensor then acts as a reversible hydrogen electrode and when the pH and the hydrogen partial pressure are constant, it can be used as a reference electrode.

Experimental procedure

The experimental scope included the measurements of electrochemical potentials, both corrosion and electrode potentials, and the performance of

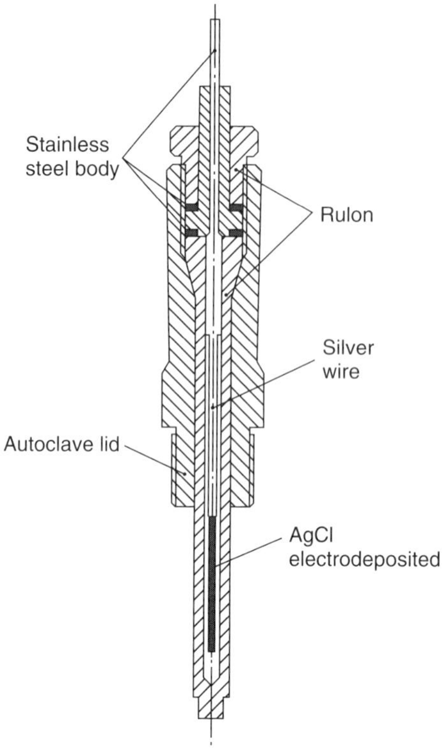

Fig. 4.2 Ag/AgCl reference electrode.

constant extension rate tests (CERT). Tests were performed in experimental facilities which simulate BWR operating conditions with high pressure–high temperature water (8 MPa and 280°C). Figure 4.3 shows a scheme of the recirculation loop and a photograph of the autoclave with four reference electrodes, used for electrochemical measurements. Experimental facilities include three independent loops suitable for the performance of stress corrosion cracking (SCC) tests. The autoclave, used for electrochemical measurements, is located in an independent by-pass close to any of the autoclaves for SCC testing.

The experimental facilities are made up of a water-treatment system and a high pressure–high temperature loop. The final product of the water-conditioning system is high purity water with $0.056\,\mu S\,cm^{-1}$ conductivity. Deionized water is deaerated by vigorous purging with argon gas, through a series of three connected tanks. The make-up tanks are filled with test solutions, with the different impurities, bubbled and pressurized with hydrogen or oxygen in order to produce the required concentrations. Dissolved gas concentration and conductivity are automatically and continuously

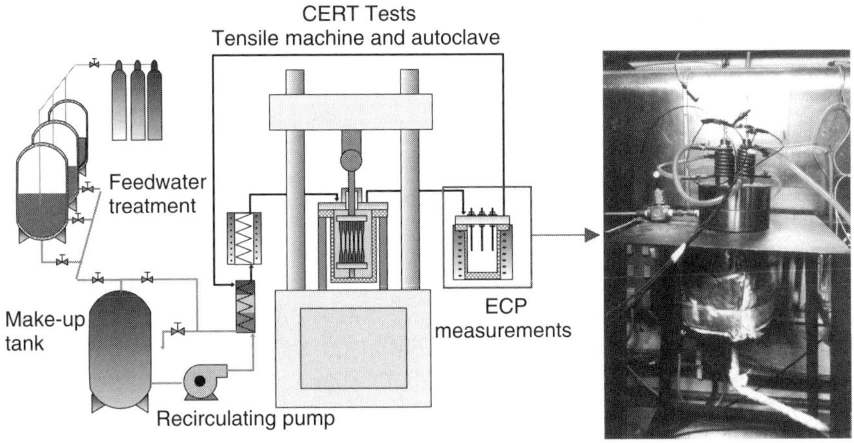

Fig. 4.3 Boiling water reactor experimental facilities (CERT = constant extension rate tests; ECP = electrochemical corrosion potential).

measured. Solutions are recirculated from the make-up tanks, through the autoclave, and returned to the make-up tanks. The make-up tanks and drain solutions are sampled and analyzed at each start-up, periodically during testing and at the end of the test.

The recirculation loop used in this experimental work includes two autoclaves. One autoclave is coupled with a mechanical system to perform CERT on six specimens. Another one-litre stainless steel autoclave is located close to the first autoclave. Four different reference electrodes can be fitted and tested simultaneously in this autoclave.

Laboratory studies have been carried out with two baseline BWR environments: normal water chemistry (NWC) with 200 ppb of dissolved oxygen; and hydrogen water chemistry (HWC) with 125 ppb H_2 + 20 ppb O_2. Different types of impurities and concentrations were added in order to evaluate the effect on the corrosion tests and the ECP measurements. Chemical additives used in this study included Na_2SO_4, ClNa and $CuCl_2$. Usually, only one anion was added in each test with different concentrations from 20 ppb up to 100 ppb. The addition of these impurities, within the ppb range, did not change the solution pH and the variation of water conductivity was found to be no more than 1 $\mu S\,cm^{-1}$.

The behaviour of reference electrodes was checked by means of electrode potential and corrosion potential measurements. Electrode potentials were verified easily under HWC, using a platinum electrode. Such measurements were used to assess the viability and behaviour of the sensors, comparing theoretical and experimental values. Corrosion potential measurements were carried out for different materials: austenitic stainless steels, type AISI 304 and 316NG, and nickel-based alloys. Comparisons between the corrosion potentials, obtained with the different electrodes, were made in order to study the reliability and to confirm the effect of impurities.

Results

Measurements of corrosion potential were equal and reproducible. Results obtained with these measurements were consistent with those reported in the literature data. The corrosion potential showed a clear dependence on the dissolved oxygen concentration. The changes observed in corrosion potential due to the presence of impurities were small and in practice sometimes negligible due to the small conductivity changes [5]. Here, only the most representative measurements are included. The potential measurements were converted to the SHE scale, using the theoretical values calculated for each electrode, and these values are shown in the following figures.

Figure 4.4 shows the corrosion potential measurements of Inconel 600 and AISI 316NG SS, under NWC with 20 ppb Cu^{2+}, measured against two different reference electrodes. The reference electrodes used were the internal Ag/AgCl electrode and YSZ electrode, with Cu/Cu_2O mixture. An

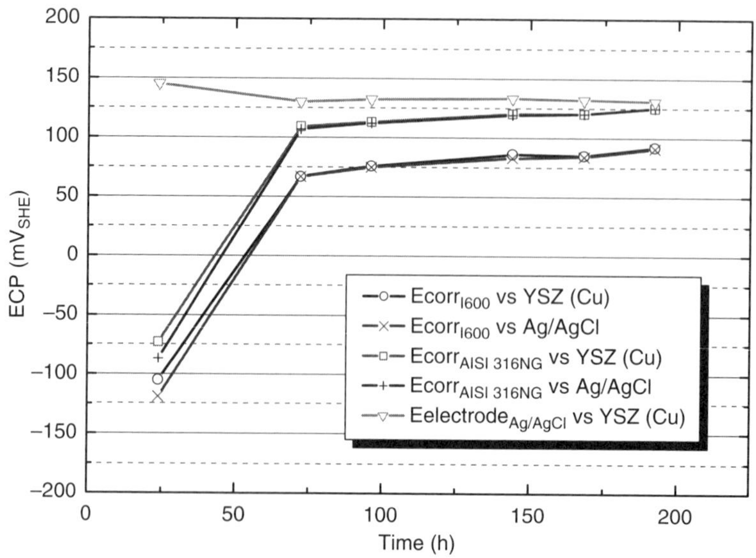

Fig. 4.4 Electrochemical corrosion potential (ECP) and Ag/AgCl electrode potential, under normal water chemistry + 20 ppb Cu^{2+}.

increase of corrosion potential, probably owing to oxide film formation, was observed in the first days when the electrode surface was fresh. Corrosion potential measurements of each material were quite similar, independent of the reference electrode used. In addition, the Ag/AgCl electrode potential measured against the YSZ (Cu) electrode was close to the theoretical value.

The corrosion potential of AISI 304 SS, measured against three different reference electrodes (Pt, YSZ and Ag/AgCl), under HWC environment plus 100 ppb SO_4^{2-}, can be observed in Fig. 4.5. During the first days, the corrosion potential decreased until the desired gas concentration was achieved. The same value of corrosion potential was obtained with YSZ (Cu) and Pt electrodes. The use of dissolved hydrogen allows the use of the Pt electrode as a hydrogen reference electrode. This electrode presents very good behaviour and reproducibility with this environment and is a good reference in order to compare the behaviour of the other electrodes. YSZ electrode potential measured against Pt electrode showed a good response and was close to the theoretical value. In addition, the electrode potential remained constant when the hydrogen concentration was increased.

However, Ag/AgCl electrode potential was clearly lower than the theoretical value (130 mV_{SHE}). An electrode potential decrease and a corrosion potential increase can be observed during the course of the entire test with dissolved hydrogen when the potentials measured against this electrode are compared with other electrodes. This effect was higher when hydrogen

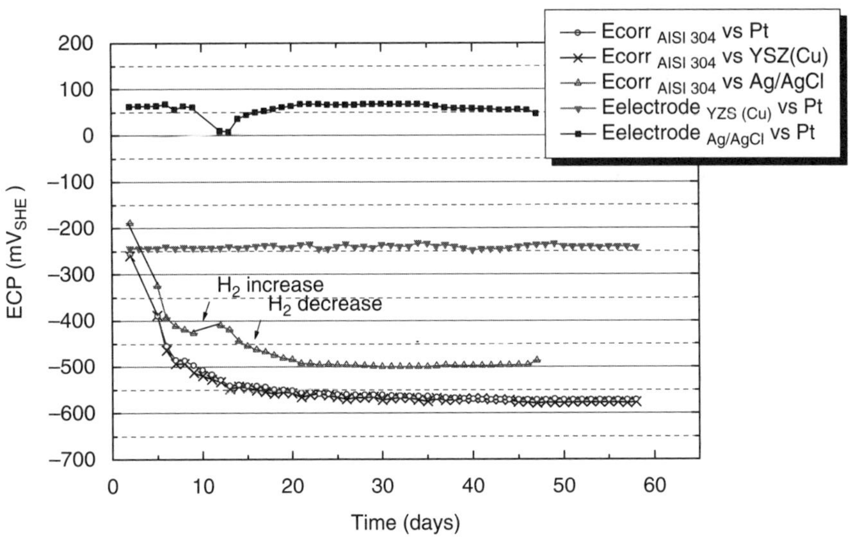

Fig. 4.5 Corrosion and electrode potentials under hydrogen water chemistry + 100 ppb SO_4^{2-}.

concentration was increased (Fig. 4.5). The decrease in the electrode potential produced a shift in the corrosion potential measurements against this electrode. The electrode potential behaviour is caused by a secondary reaction due to moderate concentrations of hydrogen diffusing through the Rulon wall [6].

From all tests performed under BWR operating conditions, the following results were obtained for each electrode:

1 The platinum electrode showed a reversible behaviour in chemistry with hydrogen addition (HWC). However, some failures have been detected with high levels of oxygen or with the presence of electroactive species. This is the case of copper ions that can be reduced easily over the platinum surface with dissolved hydrogen.
2 The Ag/AgCl internal electrode showed some problems concerning the accuracy of electrode potential when dissolved hydrogen was used, as mentioned above. In addition, special care must be taken with the design in order to avoid an additional potential drop due to isothermal liquid junction potential. The design of Ag/AgCl external electrodes enables measurement of this potential with an electrode potential calibration at room temperature. Most of the internal Ag/AgCl electrodes use demineralized water inside, and the electrode potential is a function of the silver chloride solubility with temperature. The electrode potential could be influenced by the liquid junction potential and it is not possible to calibrate it.
3 The YSZ electrode has shown a reversible behaviour, good stability and reproducibility. The best characteristic of this electrode is its easy and reproducible construction. With regard to the chemical and thermodynamic viability of this electrode, the thermodynamics of the YSZ membrane have been extensively studied by different authors [7, 8]. The behaviour of the Cu/Cu_2O mixture was checked using electron spectroscopy for chemical analyses (ESCA). The mixture was analyzed, before and after the electrode use, and in no case was Cu(II) found, indicating a reversible behaviour.

4.2.2 Tests under pressurized water reactor operating conditions

The tests performed under simulated PWR operating conditions at CIEMAT are enclosed in the round robin tests of the LIRES project. The objective of the LIRES project is to develop a reliable and robust high temperature reference electrode for in-core electrochemical measurements in light water reactors (LWR). Four different high temperature reference electrodes have been developed and the proper working of these electrodes has been studied in a round robin test. Special attention is paid to the design and the testing of the proper working of the reference electrodes under PWR requirements and irradiation conditions.

High temperature reference electrodes

The validity of three electrodes, which have been developed under the LIRES project, has been evaluated in accordance with the test procedure of the round robin. Reference electrodes have been described in detail in previous reports [9, 10] and in other chapters in this volume. The following high temperature reference electrodes were tested at CIEMAT laboratories:

YSZ electrode with a Ni/NiO mixture

This electrode consists of an yttria-stabilized zirconia ceramic tube, filled with an internal junction of nickel and nickel oxide (Ni/NiO) solid mixture. The electrochemical behaviour is similar to the electrode mentioned above with a powder mixture of Cu/Cu_2O. The advantage of this electrode is the innovative metal–ceramic sealing by magnetic compression which allows the electrode to be used at higher temperatures. The YSZ electrode has been developed by SCK·CEN under work-package 2.1 of the LIRES project [11].

Palladium hydride electrode

A simplified design version of the palladium hydride electrode was used in the round robin tests of the LIRES project. The electrode consists of two small wires of palladium and platinum welded to two long wires of stainless steel covered with heat-shrinkable Teflon. This Teflon tube provides an electrical insulation although its use limits the maximum temperature to around 300°C. The palladium electrode has been developed by VTT under work-package 2.4 of the LIRES project [12].

The palladium electrode is polarized cathodically against the platinum electrode. Therefore, a constant current (50 μA) is applied which produces a constant hydrogen concentration at the palladium electrode surface and it can act as a hydrogen reference electrode. The electrode potential is a function of the pH and hydrogen concentration, so it can be used as a reference electrode when both parameters are constant.

External Ag/AgCl electrode

The external Ag/AgCl electrode consists of a silver wire with a silver chloride layer deposited. This redox couple was connected to the test solution by means of a salt bridge made of a sapphire tube with an asbestos wick filled with distilled water. Sapphire tubes were used as materials for the high temperature parts because of the stability and lifetime in boric acid. The electrode potential depends on the Cl^- concentration reached by the maximum solubility of AgCl. It is necessary to measure the electrode potential at room temperature and a correction due to thermal liquid junction potential is required. The Ag/AgCl electrode has been developed by NRI under work-package 2.2 of the LIRES project [13].

Platinum electrode

This electrode consists of a small wire of platinum point-welded to a zirconium wire. This combination is oxidized in an oven at 500°C. This thermal treatment produces a layer of zirconium oxide which isolates the conducting zirconium wire. This procedure allows the use of this electrode at higher temperatures than electrodes isolated with Teflon.

In addition, a commercial Ag/AgCl electrode has been introduced in order to have another reference which is not affected by the hydrogen or oxygen concentration. This electrode is an external Ag/AgCl electrode with a salt bridge comprising a Teflon tube filled with asbestos. The working is similar to that of the LIRES external Ag/AgCl electrode but the use of the Teflon tube limits the maximum temperature range to around 300°C.

Experimental procedure

The experimental procedure includes tests at room and high temperatures in order to characterize not only the reference electrodes but also the measurement system. However, the most relevant results are related to the study of the electrodes' behaviour during the functional testing with different high temperatures, the effect of an oxygen trip and the lifetime testing. On the functional testing, the accuracy and reliability of these electrodes have been studied by means of electrode potential measurements under typical PWR conditions.

Functional testing at high temperature consists of the measurements of electrode potential against ground (autoclave body) and with the Pt electrode used as reference electrode. The potential measurements were carried out using a Keithley multimeter model 2700 with a data acquisition system. This system provides different channels of multiplexed measurement. During the test, several parameters are recorded using a personal computer, such as temperature, electrode potentials, dissolved gas concentration and conductivity.

High temperature tests were performed in high temperature–high pressure facilities, suitable for simulating PWR conditions. PWR facilities consist

Fig. 4.6 Pressurized water reactor experimental facilities.

of a refreshed autoclave feed by independent high temperature–high pressure loop (Fig. 4.6).

The water-conditioning system has a prefiltration by a carbon filter and a reverse osmosis system followed by deionization (model Milli-RX system). The ultrapure water quality is obtained by the Milli-Q 185 Plus system, in which the water is exposed to ultraviolet light and a water purification pack (Q-PAK). The final product is high purity water with $0.056\,\mu S\,cm^{-1}$ conductivity.

Deionized water is deaerated by vigorous purging with argon gas, through a series of three connected tanks. Solutions containing boric acid and lithium hydroxide are deaerated by purging with argon and added to the pure water. Test solutions are analyzed to ensure the desired range. The make-up tanks are filled with test solutions, bubbled and pressurized with hydrogen or oxygen in order to produce the required level. Dissolved gas concentration and conductivity are automatically and continuously measured. Solutions are recirculated from the make-up tanks, through the autoclave, and returned to the make-up tanks.

The autoclave is made of AISI-316 with a capacity of 8 l. Reference electrodes were introduced by four holes in the autoclave cover. Figure 4.6 shows the autoclave with the LIRES reference electrodes. Inside the autoclave the active tips of the reference electrodes were located as close as possible, ensuring that no electrical contact could be produced between them.

All the tests were performed with a solution of 1200 ppm of boric acid and 2 ppm of lithium hydroxide. The dissolved gas concentration used was

2 ppm of hydrogen, except for the oxygen trip test with 300 ppb of dissolved oxygen.

Results

As mentioned above, the functional testing included higher temperatures than 300°C, lifetime testing and oxygen trip. Here, only the results of two of them are shown in order to assess the behaviour of each reference electrode. The most important changes in the response of reference electrodes have been observed depending on the dissolved gas used. This effect is shown in the tests with hydrogen and the oxygen trip, both at 300°C.

Test at 300°C with dissolved hydrogen

Figure 4.7 shows the electrode potential measurements, obtained at 300°C, in water with 1200 ppm of boric acid and 2 ppm of lithium hydroxide and 2 ppm of dissolved hydrogen. The electrode potential of the Ag/AgCl electrode showed a high level of noise at the beginning of the test, probably due to the presence of small bubbles inside the sapphire tube. After 20 h of testing, the scatter decreased because the bubbles were reduced by the effect of pressure and temperature. This is a typical behaviour of Ag/AgCl

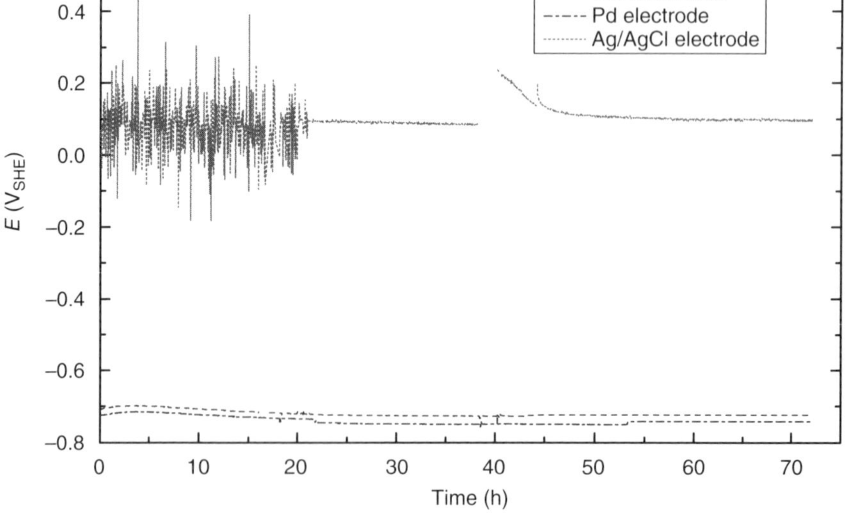

Fig. 4.7 Electrode potentials obtained at 300°C with 2 ppm of dissolved hydrogen.

electrodes if bubble size is very small and the conductivity through the tube is not interrupted.

The YSZ and palladium electrode responses are clearly observed in Fig. 4.8 using different potential scale drawing. Regarding the YSZ electrode, when the electrode potential was stabilized, after around 20 h, the electrode working was correct with a potential value close to the theoretical value ($-730\,mV_{SHE}$).

During the test, the polarization of the palladium electrode was changed from 10 to 50 μA in order to study the behaviour of the electrode. When the applied current is at the highest value, the potential is close to the theoretical value ($-744\,mV_{SHE}$) and it indicates that the electrode is working correctly as a hydrogen reference electrode.

At the end of the test, the resistance measurement of the YSZ electrode showed a decrease at high temperature, in spite of the proper working of the electrode. At room temperature, the resistance of the YSZ electrode remained lower than the value before the test. A careful examination of electrodes was performed, when the running was finished, and no damage to isolated electrical wires or the ceramic membrane was detected.

Examination of the Ag/AgCl electrode revealed a large number of big bubbles inside the sapphire tube that could interrupt the conductivity through the tube. Bubbles were carefully removed and a new asbestos wick

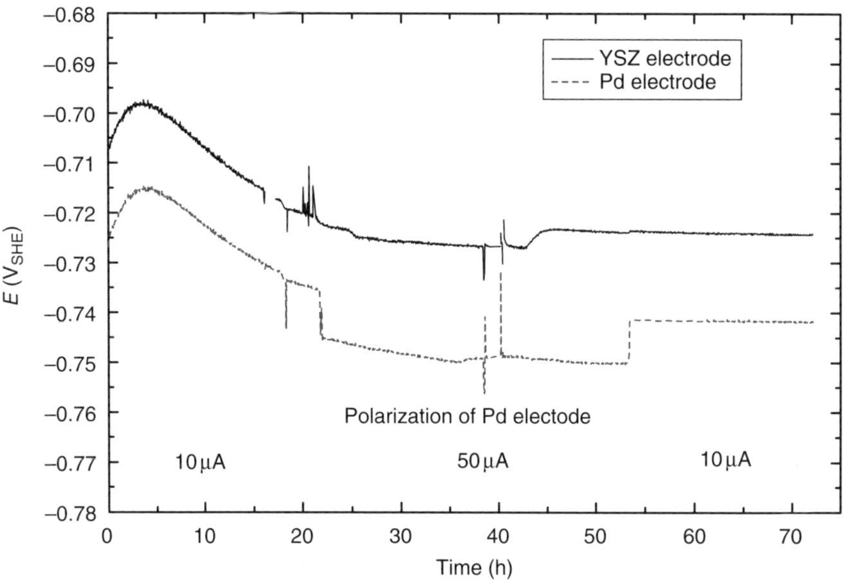

Fig. 4.8 Yttrium-stabilized zirconia (YSZ) and Pd electrode potentials obtained at 300°C with 2 ppm of dissolved hydrogen.

was used. After this procedure, the electrode potential was again compared with SHE at room temperature and a small decrease was observed (338 instead of 420 mV_{SHE}). However, this difference did not affect the correct behaviour of the reference electrode at high temperature in the following tests.

Test at 300°C with 300 ppb of dissolved oxygen

The test was performed using oxidizing conditions (300 ppb of dissolved oxygen) followed by reducing ones (2 ppm of dissolved hydrogen). It has been considered that the desired gas concentration is achieved correctly when the difference between the inlet and outlet gas concentrations is less than 25% of the inlet concentration. Previously to the test, a preconditioning of the recirculation loop was carried out in order to begin the test with the desired oxidizing conditions when the consumption of the oxygen into the loop was stabilized.

The water chemistry and temperature are shown in Fig. 4.9. During the course of 143 h, the inlet oxygen concentration of 300 ppb was maintained. Measurement of oxygen level at the outlet water showed that only a small amount of oxygen was consumed and the concentration was higher than 250 ppb O_2. Prior to the change to reducing conditions, an intermediate step

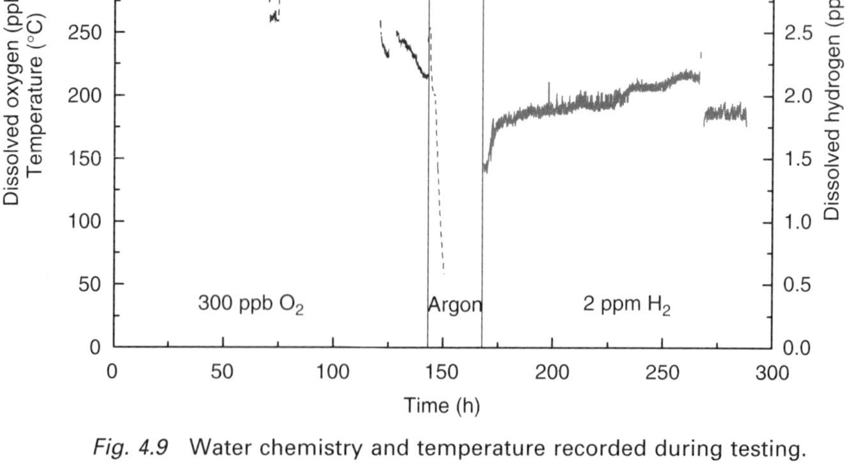

Fig. 4.9 Water chemistry and temperature recorded during testing.

with argon gas was performed in order to eliminate all the dissolved oxygen (26 h). Finally, dissolved hydrogen was introduced for 129 h.

The potential measurements are plotted in Fig 4.10, using as reference the Ag/AgCl electrode developed under the LIRES project. All of the potentials showed the same decrease at the middle of the test due to a short malfunctioning of this reference electrode. The other Ag/AgCl reference electrode presented a constant electrode potential in spite of the gas added. However, this value was lower than the theoretical one and a large scatter can be observed. This behaviour is due to a large number of bubbles inside the Teflon tube, as confirmed at the end of the test.

The potential of the YSZ electrode was quite stable and close to its theoretical value. Electrode potential has not been influenced by the changes in the gas concentrations. This electrode presented good behaviour as a reference electrode with the electrode potential independent of the environment when the pH was maintained at the same value.

Reference electrodes of noble metal, both palladium and platinum electrodes, were strongly influenced by the presence of dissolved gas, as expected. The platinum electrode presented a stable potential with oxidizing conditions and this value changed with reducing conditions. However, this electrode had not achieved the reversible behaviour of the hydrogen reference

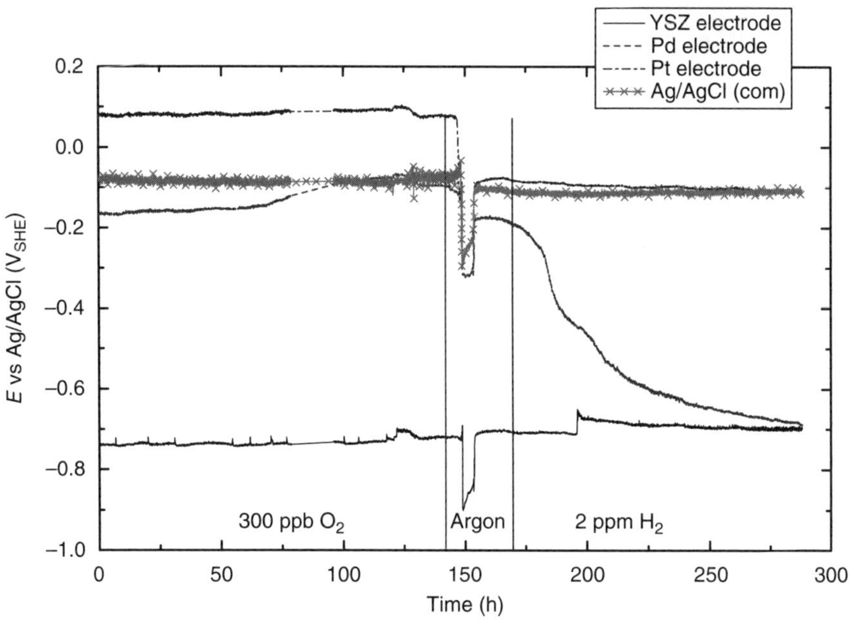

Fig. 4.10 Electrode potential measurement (standard hydrogen electrode, SHE, scale) against Ag/AgCl electrode.

electrode because the electrode potential value was quite different from the theoretical one. On the contrary, the potential of the palladium electrode at the end of the test was close to the theoretical value, indicating its correct working as a hydrogen reference electrode. The superior response of the palladium electrode could be due to the effect of electrode polarization.

From all of the tests performed under PWR operating conditions, the following results were obtained for each electrode:

External Ag/AgCl electrode

Most of the electrode potential measurements were stable and reliable. Only at the beginning of the tests was a large scatter observed due to the presence of small bubbles inside the sapphire tube. When the size of the bubbles is small, the good behaviour of the electrode is recovered at high pressure and temperature. However, when the temperature is decreased, these bubbles can grow and the conductivity through the tube can be interrupted, damaging the reference electrode.

The main characteristic of the electrode design is the use of a sapphire tube as a salt bridge. The advantages of the design with sapphire are that it facilitates the removal of bubbles and it allows use of the electrode at temperatures higher than 300°C, due to the absence of Teflon. However, the fragility of the sapphire tube imposes limitations on its use. Nevertheless, the easy handling of the electrode makes it ideal to employ in laboratory measurements.

Palladium hydride electrode

The palladium electrode worked correctly as a hydrogen reference electrode when it was tested under reducing conditions. The electrode did not work properly in water with dissolved oxygen in spite of the polarization procedure. However, it seems that polarization improved its behaviour with respect to the platinum electrode without any polarization.

The palladium electrode used in the round robin test is a simple version of the design with two wires of palladium and platinum with heat-shrinkable Teflon. The electrode examination revealed that the Teflon was damaged at the end of the tests at temperatures higher than 300°C.

YSZ electrode

The results showed that in most cases electrode potentials were stable and close to the theoretical values, giving reliable data. Measurements of the electrode resistance were somewhat low, but surprisingly these low values did not seem to affect the electrode response. Only at the end of the tests could a small deviation of the theoretical electrode potential value indicate a small deterioration of the electrode. This behaviour is probably due to the extensive employment of this electrode, which has been used for 3300 h and has suffered from several thermal cycles. This design, especially the innovative metal ceramic connection, showed a mechanical stability without any problems detected in the two different electrodes used.

4.3 Conclusions

A comparative analysis of the electrode behaviour differences, based on CIEMAT experience, has been carried out taking into account the chemistry of the environment, the temperature, the materials used and the design.

Regarding the design, the high temperature reference electrodes could be categorized in three main types:

1 Ag/AgCl electrodes: The first advantage of this kind of electrode is that the electrode potential is only a function of the chloride activity. This reference electrode is not influenced theoretically by the environment tested, like pH or gas concentration. Therefore this is the only high temperature reference electrode that should be used on tests where a change in the pH of the solution can be expected.

These electrodes can be divided into two classes: internal or external. The advantage of the internal reference electrodes is that the electroactive element is maintained at the same pressure and temperature as the system, avoiding thermal liquid junction potential. The disadvantage of this electrode is that sometimes it can suffer AgCl reduction by the diffusion of dissolved hydrogen through the electrode walls and also thermal decomposition. This effect produces a shift of the electrode potential and higher corrosion potential measurements.

The use of a salt bridge on the external electrodes produced a thermal liquid junction potential across the electrolyte bridge, although its accuracy is enough for the corrosion studies. On the contrary, this design avoids contamination of the electroactive sensor. The main disadvantage of this type of sensor is that there are some problems associated with the presence of bubbles inside the salt bridge. As demonstrated previously, sometimes when the bubbles were small and the conductivity was not totally interrupted, the proper working of the electrode was recovered and the scatter decreased. Special attention must be placed on the effect of the formation of bubbles when several thermal cycles are used. Another design includes a valve that relieves gas bubbles, thus avoiding the problem, although it limits the use to laboratory measurements [14].

2 Noble metal electrodes: Several types of noble metal electrodes are available. The simplest is the platinum electrode which presents a high reproducibility as a hydrogen reference electrode. However, this reference electrode is strongly influenced by the presence of oxidizers, especially small concentrations of dissolved oxygen or species that could be easily reduced over its surface, such as copper ions. As mentioned above, the palladium hydride electrode seems to present a better response due to the polarization of the electrode.

3 YSZ electrodes: YSZ electrodes have shown good behaviour independent of the powder mixture of metal/metal oxide used. The only requirement is that the stoichiometric of the internal couple is well defined. The main advantages are that a calibration electrode is not necessary and the ceramic membrane prevents the contamination with other species in the environment. However, its use is limited to measurements where pH is constant and known.

Taking into account the limitations on the use of the reference electrodes mentioned above, the choice of the better high temperature reference electrode can depend on different aspects of the measurement as follows:

1 Location of measurements: The most important requirement of the reference electrode can be different if the measurements are to be performed in the laboratory or in-core measurements in LWRs. In the first case, a reference electrode with easy assembly and operation is enough, as in Ag/AgCl electrodes. However, for in-core measurements the reference electrode must have a robust design and the system integrity must be assured for a long time under high temperature, high pressure and high irradiation conditions.

2 Operating temperature: The main effect of the temperature is based on the materials used in the electrode design. When operating temperature is maintained around 300°C (as under BWR-simulated conditions), the design of high temperature reference electrodes is facilitated by the use of plastic materials such as Teflon or Rulon. These materials allow the electrical isolation and the seal with an easy and simple design.

  The measurements under PWR conditions with higher temperatures make it necessary to improve the design. These high temperatures are just over the range of use of plastic materials and it is only possible to use metallic or ceramic materials. These kinds of materials may further complicate the electrode design. Thus there is much interest in the innovative metal ceramic connection of the YSZ electrode developed under the LIRES project which shows very good mechanical stability.

3 Chemical environment: Regarding the chemical environment, several aspects must be considered: dissolved gas concentration, pH and other electroactive species. For example, the presence of other electroactive species dissolved in the environment limited the choice of the reference electrode to YSZ or external Ag/AgCl electrodes. Only the design of these two electrodes avoids the electrochemical sensor being contaminated by other redox reactions.

The use of noble metal electrodes as reference electrodes is obviously limited to the presence of dissolved hydrogen. Thus they can be used with a very good response in a PWR environment with hydrogen. But also, one of the most important applications can be the control of HWC because it is very sensitive to the presence of small amounts of oxidizing conditions. However, when it is important that the changing of dissolved gas concentration does not affect the electrode potential, the best behaviour has been obtained with the YSZ electrode.

The pH of the environment affects the electrode potential of both noble metal and YSZ electrodes. So, if the pH is changed during the measurement, only the Ag/AgCl electrode can be used. However, if the pH remains constant (as for example in BWR conditions), it is possible to use any of the reference electrodes. Measurements with several reference electrodes at the same time allow using one of them as a pH sensor.

4.4 References

1. D. D. Macdonald, *Corrosion*, 1978, **34**, 3.
2. J. H. Zheng, W. Bogaerts. *Reference Electrodes for Electrochemistry and Corrosion Study in High Temperature Aqueous Environments.* LIRES project, May 2001.
3. M. Navas, D. Gómez Briceño. *Nucl. Engng Design*, 1997, **168**, 183.
4. L. W. Niedrach. *J. Electrochem. Soc.*, 1982, **129**, 1446.
5. C. C. Lin, F. R. Smith, N. Ichikawa, M. Itow. *Corrosion*, 1992, **48**, 16.
6. M. Indig. *Corrosion*, 1990, **46**, 680.
7. D. D. Macdonald, S. Hettiarachchi, S. J. Lenhart. NP-6005, EPRI, Palo Alto, CA (1990).
8. L. W. Niedrach, W. Stoddard. *Corrosion*, 1985, **41**, 45.
9. R.-W. Bosch, *et al. Int. Conf. Environmental Degradation of Materials in Nuclear Power Systems – Water Reactor*, Stevenson, USA, 2003.
10. R.-W. Bosch, G. Nagy, D. Féron, M. Navas, W. Bogaerts, D. Kárník, T. Dorsch, A. Molander, K. Mäkelä. EUROCORR 2004, Session 46: Nuclear corrosion, Paper 46-O-334, Nice, France, September 2004.
11. R.-W. Bosch, R. Nieuwenhove. Development of mixed metal oxide high temperature reference electrode. LIRES work-package 2.1. Report R-3650 (2002).

12. K. Makela, T. Saario. Development of LWR reference electrodes. Preparation of reference electrode IV. LIRES work-package 2.4. Research report NO BTUO76-021075 (2002).
13. D. Karnik, M. Paulovic, J. Kocik, E. Keilova. Design and development of reference electrode II, Ag/AgCl (Work-package 2.2). NRI Report (2002).
14. D. D. Macdonald, A. C. C. Scout, P. Wentrcek. *J. Electrochem. Soc.*, 1979, **126**, 1618.

Part II

Electrochemistry and corrosion issues

5

The influence of corrosion potential on stress corrosion cracking of stainless steels in pressurized water reactor primary coolant environment

M. POSTLER, NRI Řež, Czech Republic;
R.-W. BOSCH and S. VAN DYCK, SCK·CEN, Belgium

5.1 Introduction

Due to ageing of the nuclear power plants, stainless steel core components are subjected to increasing radiation doses. This exposure enhances their susceptibility to a special type of stress corrosion cracking (SCC), known as irradiation-assisted SCC (IASCC), which is influenced by high temperature water environment, mechanical stresses and the presence of irradiation. One way to assess IASCC is to measure the corrosion potential, which estimates whether it is likely that IASCC occurs (high value of potential) or not (low value).

The main task of this research project is to try to find a direct link between the value of the corrosion potential and SCC present in stainless steels exposed to pressurized water reactor (PWR) conditions. Different values of the corrosion potential have been achieved by changing the water chemistry environment by adding oxygen or nitrogen into the autoclave.

5.2 Experiment

All tests were performed with AISI304. The chemical composition is shown in Table 5.1, together with the mechanical properties tested at 288°C in an inert atmosphere. The steel was cold rolled by 35%.

Two sets of samples were prepared for the experiment. For the electrochemical measurements and analyses of oxide layers, flat samples $10 \times 10 \times 3$ mm were prepared. For evaluation of mechanical behavior, standard slow strain rate tests were used.

All samples were exposed in the autoclave to simulated PWR primary circuit conditions (1000 ppm B, 2 ppm Li, 27 ppm H_2, temperature 300°C,

Table 5.1 Chemical composition (weight %) and mechanical properties of tested material

C	Cr	Ni	Mn	Mo	Si	S
0.04	18.2	8.7	1.22	0.53	0.33	<0.001
YS_{288} = 650 MPa	TS_{288} = 650 MPa	HV = 290				

Table 5.2 Solution for electrochemical etching

Water	330 ml
Phosphoric acid (85%)	550 ml
Sulfuric acid	120 ml

pressure 140 bar). There were basically three variations of the environmental conditions; the first is the above-mentioned standard PWR, identified as hydrogen in the following text. The second type of environment consisted of standard PWR conditions with 400 ppb of oxygen injected into the autoclave; this condition is identified as oxygen in the following text. The last is similar to oxygen except that the injected gas was nitrogen; this is identified as the nitrogen condition. The exposure of all tests was approximately 10 days.

5.2.1 Flat samples

A stainless steel wire was spot-welded to each of the flat samples. This wire was then covered with PTFE tube in order to isolate it from the autoclave environment. Previous experiments showed that if the temperature does not exceed 300°C, this protection is sufficient.

The complete procedure of preparation of the sample surface consisted of several steps: first, both sides of all flat samples were ground on the silicon carbide papers in sequence 320–600–1200. In order to reveal the true microstructure of the metal on the sample surface and to remove all artifacts produced by mechanical grinding (disturbed metal, mechanical twins and scratches), one side of each sample was electropolished. The method described in ASTM E1558-93 was used. Several solutions and parameters of electropolishing were tested and each time the effect on the surface was observed under a light microscope. It appears that higher contents of H_2SO_4 may create intergranular attack on the sample surface; therefore it was decided to use in addition phosphoric acid. The composition of the solution used for polishing is shown in Table 5.2.

The test standard proposes current densities of $0.05\,A\,cm^{-2}$ during a period of 1 min. No information is given about the potential, which should be applied. With respect to the surface of the sample, the current was calculated to be approximately 100 mA, resulting in a potential of about 1.8 V. The polishing time of 1 min proved to be too long and after several tests it was established that the ideal time for polishing is approximately 30 s. No film formation is assumed to take place due to the low pH. The surface bears no traces of corrosion attack and is considerably smoother compared to grinding. All samples were finally cleaned in the ultrasonic bath in acetone and ethanol.

X-ray photoelectron spectroscopy (XPS) measurements were performed on a PHI 1600 photoelectron spectrometer equipped with an Omni Focus Lens III using a standard Al Kα X-ray source (15 kV, 350 W) and the hemispheric electron energy analyzer with maximum resolution 0.8 eV. The analyzed area had a diameter of 0.8 mm. Spectra were obtained at a take-off angle of 45° relative to the sample surface. In order to obtain depth profiles, the oxide layer was gradually removed by ion sputtering. The surface sputtering was performed using 4.0 kV argon ions, with an ion beam current of 20 mA, Ar gas pressure of 6×10^{-5} Torr and an ion gun beam spot of $3 \times 3\,mm^2$.

The depth profile is obtained by measurements that are followed by sputtering for the given period of time. This process is repeated until the oxide content is close to zero. The thickness of the layer can be estimated with reasonable precision by the total length of the sputtering. From the sputtering of a standard sample with oxide layer of known thickness it is possible to establish the correlation between length of the sputtering and thickness. For the parameters used the sputtering rate is approximately $3\,nm\,min^{-1}$. The oxide layer–bulk material interface is not calculated from the point where the oxide content reaches zero. Due to the diffusion and effects connected with ion sputtering, the boundary of the oxide layer is considered at the point where the concentration curves of Fe and O intersect in the graph.

All electrochemical measurements were performed with a Gamry potentiostat and converted to the standard hydrogen electrode (SHE) scale. As a reference electrode, an external pressure-balanced Ag/AgCl reference electrode (EPBRE) was used [1]. For comparison, part of the measurement was performed with respect to a newly developed yttrium-stabilized zirconium reference electrode (YSZRE). An artificial potential was applied using a Wenking TG97 potentiostat. Potentiodynamic scans were performed in the interval from –0.5 V vs E_{OC} to +2 V vs E_{OC} at a scan rate of $600\,mV\,h^{-1}$ (E_{OC}: open circuit potential). Since only qualitative features of the polarization curves were of interest, no potential drop compensation has been made.

5.2.2. Slow strain rate test samples

Standard specimens for slow strain rate tests (SSRTs) were manufactured with gage diameter 2.5 mm and gage length 10 mm. Two wires for electrochemical measurements were spot-welded in the area between the end of the gage and the thread on both sides of the sample. The samples were insulated from the loading system by means of ZrO-coated fittings. Secondary protection against radial contact was realized by a PTFE band.

All the tests were performed with a strain rate of $5 \times 10^{-7} s^{-1}$. After the tests the fracture surfaces were analyzed by means of scanning electron microscopy (SEM) and the main mechanical values were calculated.

5.3 Results

5.3.1 Electrochemical measurements

Results from the electrochemical measurement of the open circuit potential are shown in Fig. 5.1. One curve represents measurements against the EPBRE, the other against the YSZRE. A total of six different samples were measured, two for each condition. The initial and final parts of the curves were omitted, as they are affected by the electrochemical noise created during start-up and cool-down of the autoclave.

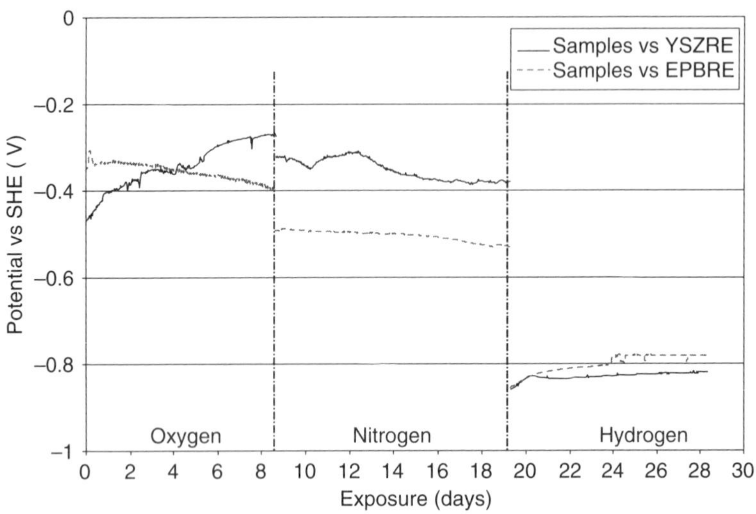

Fig. 5.1 Open potential measurements against external pressure-balanced Ag/AgCl reference electrode (EPBRE) and yttrium-stabilized zirconium reference electrode (YSZRE). SHE, standard hydrogen electrode.

It is clear that the measurements performed against the EPBRE are more stable than against the YSZRE, with one exception being the small fluctuation in the middle part of the hydrogen test. Also the average values of the E_{OC} are in good correlation with expectations, the highest potential measured for oxygen conditions, lower potential for nitrogen and the lowest for standard PWR conditions, around –800 mV vs SHE.

On the other hand the measurement results performed against YSZRE were not so stable. In particular there was no difference between the values of the open circuit potential for oxygen and nitrogen conditions. Measurements for standard PWR water chemistry – hydrogen – proved to be stable and very close to the EPBRE results.

Potentiodynamic measurements were performed to detect activation/passivation characteristics of the surface layer in different environments. The results are shown in Fig. 5.2. All measurements were performed after approximately 8 days of exposure in the different environments. All measurements were carried out against the EPBRE and converted afterwards to the SHE scale.

The results follow the pattern found in previous tests, i.e., the potential of active–passive transition for hydrogen is the lowest (about –0.8 V SHE), with those for nitrogen and oxygen higher and close to each other. The E_{OC} values obtained from the polarization curves are different from the E_{OC} values measured statically. This is due to the fact that the cathodic part of the scan influences the electrode surface and so the E_{OC} value.

The position of the polarization curves taken under three different conditions (oxygen, nitrogen and hydrogen) can be related to the statically

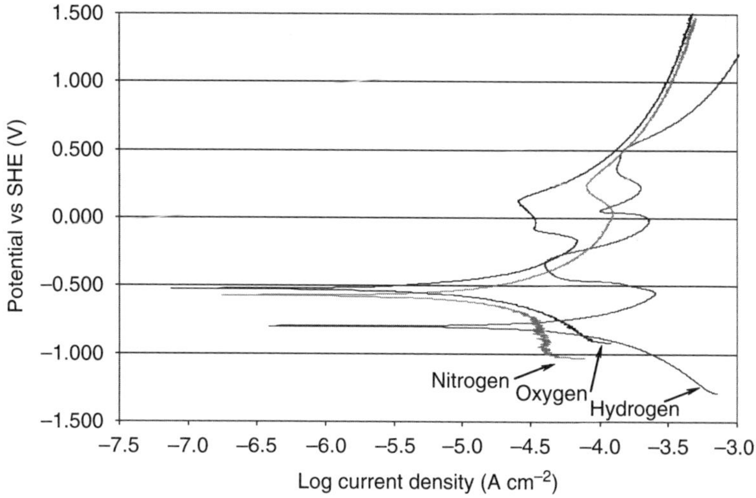

Fig. 5.2 Potentiodynamic scans. SHE, standard hydrogen electrode.

measured corrosion potentials. The polarization curve under oxygen is situated in the most anodic region followed by the curve under nitrogen. The polarization curve taken under hydrogen conditions lies in the most cathodic region. The corrosion potentials taken from the polarization curves are not exactly similar to their potentiostatically measured counterparts, as they are influenced by the cathodic part of the polarization curve. The three polarization curves all show similar behavior. Each curve has an active–passive transition and at more anodic potentials a transpassive region.

We know, however, from the results from SEM and XPS and from thermodynamics [2] that the stainless steel samples are all covered by an oxide layer. The existence of an active region in the given potential range is therefore not very likely. We assume that this typical 'active-to-passive' behavior is related to some film formation or modification that changes the polarizing capability of the stainless steel sample.

5.3.2 X-ray photoelectron spectroscopy

Unfortunately XPS is not able to distinguish between different oxides of Fe (magnetite, hematite) and also cannot detect the presence of composed oxides $(Cr, Fe)_2O_4$ and others. However, XPS can still answer the questions concerning the thickness of the oxide layer, concentration profile of the main elements and their oxidation state.

Depth profile results for the sample exposed to oxygen conditions are shown in Fig. 5.3. The elements analyzed were C, O, Cr, Fe and Ni. The concentration curves of Fe and O intersect after 64 min of sputtering; therefore the total thickness of the layer may be estimated at around 190 nm.

Carbon is present only in the first atomic layer, after the first sputtering its concentration drops sharply. Its presence is due to surface contamination and is not shown on the graph. The Ni content is unchanged through the whole thickness of the layer. Detailed analysis of XPS data shows that only the first atomic levels contain Ni in oxidized state, deeper in the layer Ni is only present in the metallic state. It corresponds with other experimental data [3] and is due to the fact that even for this environment the oxidation potential is not high enough to fully oxidize noble components of the alloy like Ni and possibly also Co.

At the outer surface layers Fe is fully oxidized; after 40 min of sputtering (approximately 120 nm, i.e., two-thirds of the layer thickness) Fe starts to appear in metallic form. The content of Cr is rather stable throughout the layer; Cr at the outer layer is fully oxidized and starts to appear in metallic form only after about 180 nm, very close to the layer–bulk interface. It corresponds to the fact that the diffusion rate for Cr in oxides is lower than that for Fe.

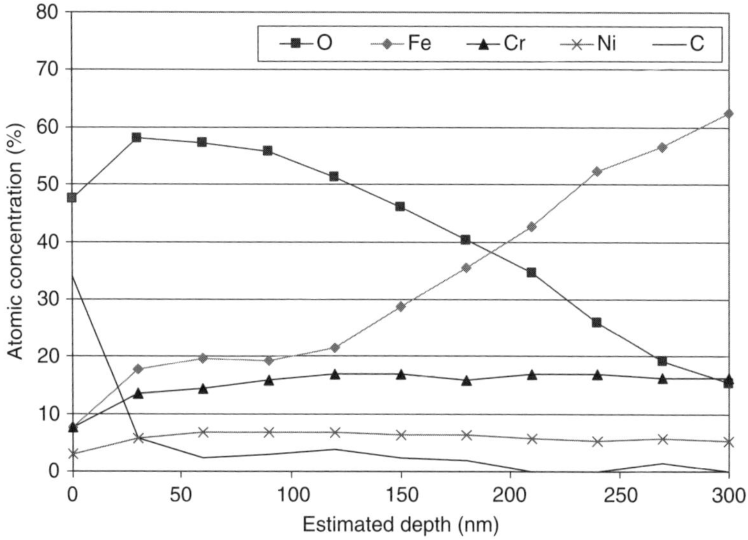

Fig. 5.3 Depth profile of the sample exposed to oxygen conditions.

Depth profile results for the sample exposed to hydrogen conditions are shown in Fig. 5.4. The elements analyzed were C, O, Cr, Fe and Ni. The concentration curves of Fe and O intersect after 50min of sputtering; so approximate total thickness may be estimated at around 150nm. This is roughly three-quarters of the thickness of the layer created during oxide conditions, which corresponds to the fact that the oxidation potential for the hydrogen conditions is lower.

Apart from the difference in thickness of the layer, composition properties are very similar to the sample tested in oxygen. Carbon is only present as surface contamination. Except for the outer surface, Ni is present in metallic form throughout the layer. Both Fe and Cr are fully oxidized at the outer surface layers. Fe starts to appear in metallic form after removing approximately 60nm of layer; Cr is partially metallic 120nm deep in the layer.

As for the exact composition of the oxide layer and mechanism of corrosion, Robertson [4] shows that corrosion of stainless steels in high temperature water is controlled by diffusion of metal ions across the oxide layer. The layer is duplex with the inner part created by oxidation of the metal surface and the outer layer created by the metal ion diffusion along the grain boundaries. Only the inner layer determines corrosion rate. The outer layer is composed mostly of Fe_3O_4; the inner layer is composed of metallic Fe, Cr and $(Fe, Cr)_3O_4$ [5–7].

The oxide is spinel rather than Cr_2O_3-based film, because the nonselective corrosion of alloy takes place during high temperature water corrosion.

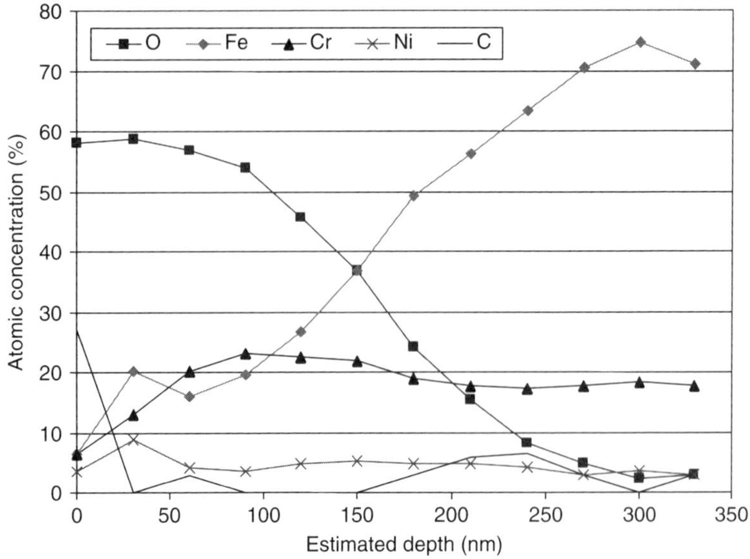

Fig. 5.4 Depth profile of the sample exposed to hydrogen conditions.

Table 5.3 Summary of results from slow strain rate tests

	Yield strength (MPa)	Tensile strength (MPa)	Total elongation (%)	Reduction of area (%)
Oxygen	621.4	640.2	15.0	56
Nitrogen	656.3	692.4	14.0	45
Hydrogen	663.1	684.7	14.0	34

Chromia film does not have sufficient time to spread after nucleation in the case of stainless steel water corrosion. The rate of nucleation falls rapidly. The model shows that with temperatures around 200–350°C it would take about 10^7 years to create a compact layer.

5.3.3 Slow strain rate tests

The results from SSRTs are shown in Table 5.3. Stress–strain curves are shown in Fig. 5.5.

The sample tested in oxygenated water (see Fig. 5.6) shows completely ductile behavior. There is no trace of transgranular SCC facets on the fracture surface; only dimples of ductile fracture are present. Uniaxial dimples of shear ductile fracture are found on the edges of the fracture surface. The

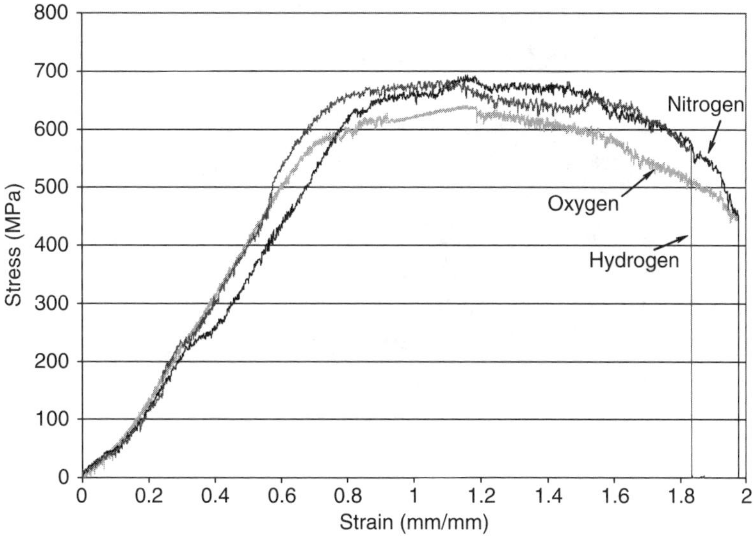

Fig. 5.5 Slow strain rate test diagrams.

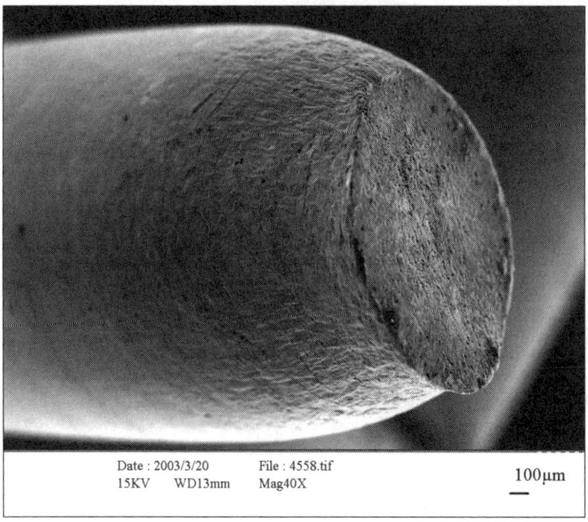

Fig. 5.6 Sample tested in oxygen conditions, general view.

plane of the fracture surface is almost perpendicular to the main axis of the sample. Also the surface of the gage (Fig. 5.7) bears no traces of brittle fracture. A high degree of ductility is also documented in the value of the reduction of area, which is more than 50%. On the other hand samples tested in both nitrogen and hydrogen exhibit the presence of SCC facets.

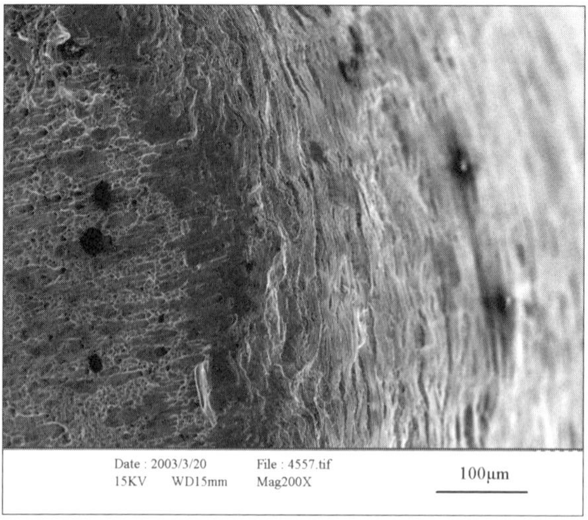

Fig. 5.7 Sample tested in oxygen, fracture surface edge and gage surface.

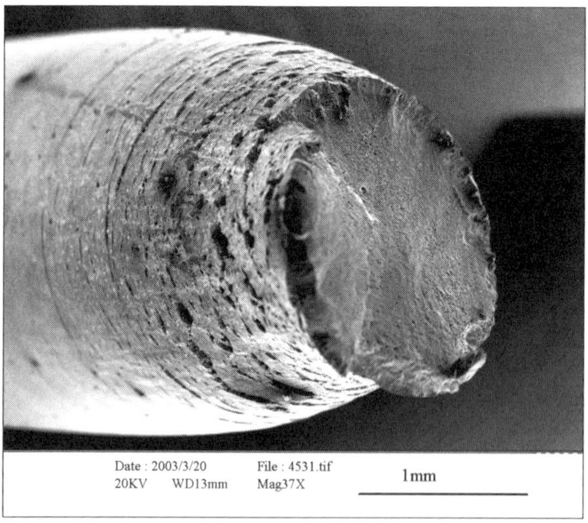

Fig. 5.8 Sample tested in nitrogen conditions, general view.

Almost the entire circumference of the sample tested in nitrogen shows traces of SCC cracks (Fig. 5.8). Locally the facets reach up to 100 μm deep. The fracture surface is inclined to the main axis of the sample. From Fig. 5.9 it can be seen that the gage surface up to a distance of approximately 1 mm from the fracture is also affected by SCC.

The ductility of the sample is reduced compared to the sample tested under oxygen. The reduction of area is 10% lower.

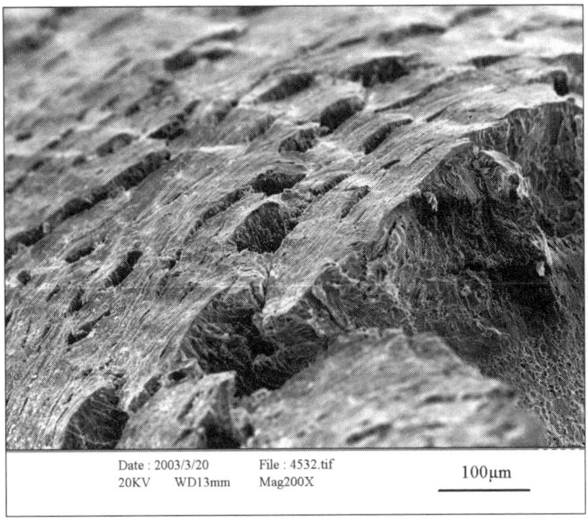

Fig. 5.9 Sample tested in nitrogen, surface of the gage.

Fig. 5.10 Sample tested in hydrogen conditions, general view.

Figures 5.10 and 5.11 show the fracture of the sample tested in hydrogen. It is clear that the fracture surface bears traces of severe SCC transgranular cracking, in some cases reaching up to 300 μm deep into the sample. SCC facets are localized and interconnected with ductile shear fracture. The fracture surface is inclined by 45° to the main axis of the sample. As in the

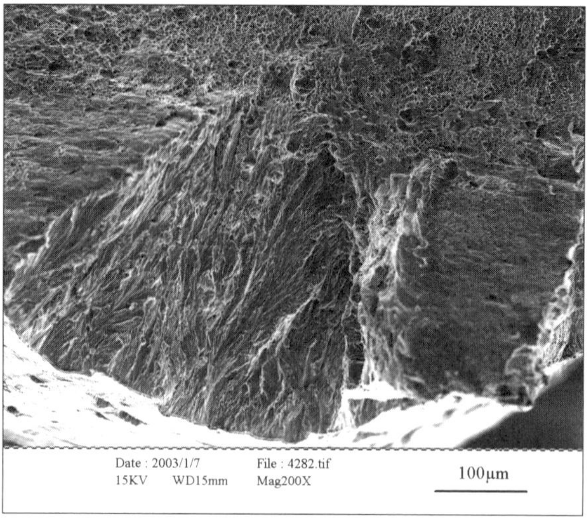

Fig. 5.11 Sample tested in hydrogen, example of stress corrosion cracking facet.

previous sample the gage surface close to the fracture is also affected by SCC.

The ductility of the sample is further reduced compared to the samples tested both in oxygen and nitrogen. The reduction of area is less than 35%.

As only one sample per condition was tested, no firm conclusions may be drawn from the mechanical tests. The results are rather close to each other, within 10%, so the differences may be caused by error of measurement. Values of total elongation are very close as well as times to failure, which are for all the samples slightly less than 130 h. This is most likely caused by the fact that even the strain rate of $5 \times 10^{-7} s^{-1}$ is relatively high and different environments do not have enough time to influence the mechanical behavior of the samples. However, by careful comparison of all data together with SEM micrographs one can see an increasing tendency towards SCC.

For the oxygen conditions there is no SCC at all and the fracture is totally ductile, with the value of reduction of area being the highest of all three samples. On the other hand the sample tested in pure PWR conditions (hydrogen) shows extended SCC facets reaching deep into the specimen. The ductility of the sample is the lowest of all three, with the RA (reduction in area) value almost half of that in oxygen conditions.

The sample tested in nitrogen is somewhere between, closer to the hydrogen conditions, both by appearance of fracture surface and by mechanical values.

5.3.4 Cracked round bars

One of the methods for the evaluation of mechanical properties of material subjected to different experimental conditions was based on tests performed on cracked round bars (CRBs). A CRB is a metal rod with a notch in the middle, from which the crack propagates to the central region.

The theoretical background of the system, the calculation of the fracture toughness for different sizes of CRBs and the evaluation of processes during pre-cracking, etc., are given elsewhere [8]. Using CRBs has several advantages. The manufacturing procedure (including pre-cracking) is well known and developed and the testing times are relatively short. On the other hand, using CRBs presents some disadvantages, mainly resulting from the fact that the crack growth region is rather limited.

An example of the resulting fracture surfaces is shown in Fig. 5.12 and 5.13 – samples tested in oxygen and hydrogen conditions respectively. The outer part of the samples is created with a notch and region of pre-cracking in air; the crack growth during the test is exhibited by rather distinctive fracture morphology in the central part of the sample; in all cases it is transgranular SCC. The sample tested in oxygen shows significantly greater crack growth, in some directions up to 400 μm, with severe secondary cracks perpendicular to the fracture plane. On the other hand the sample tested in hydrogen conditions shows a much smaller amount of cracking, less than 200 μm.

Unfortunately no reliable data were available from a.c. potential drop measurements of the crack growth rate. Even after recalculation of the

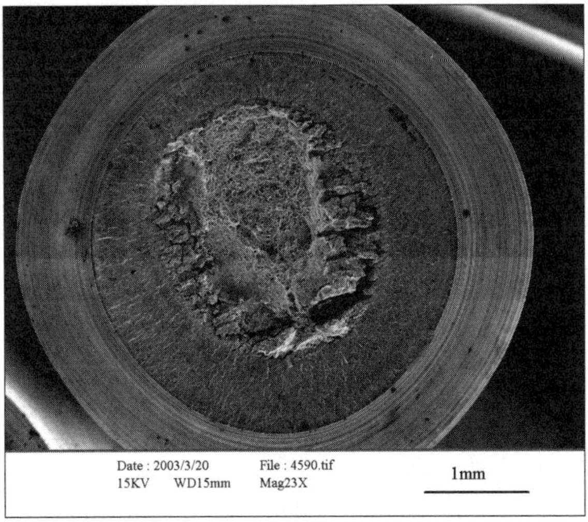

Fig. 5.12 Cracked round bar: sample tested in oxygen.

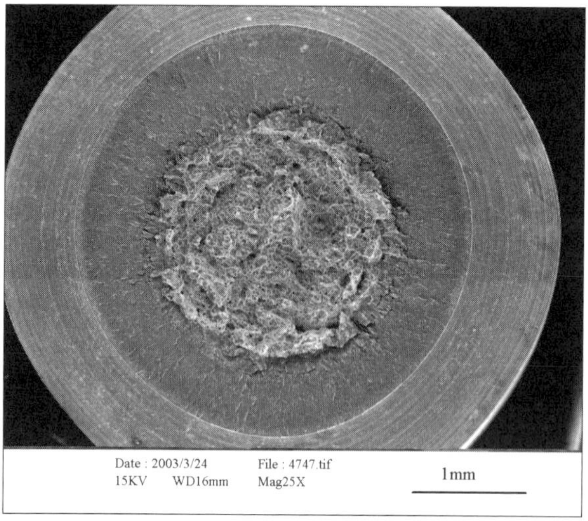

Fig. 5.13 Cracked round bar: sample tested in hydrogen.

Table 5.4 Summary of results from cracked round bar tests

	Crack increment (μm)				Average increment (μm)	Time to failure (days)	Average crack growth (mm s^{-1})	Maximum crack growth (mm s^{-1})
	Top	**Right**	**Bottom**	**Left**				
Oxygen	90	400	50	400	235	6.9	$4.0*10^{-7}$	$6.7*10^{-7}$
Nitrogen	50	50	50	50	50	9.35	$6.2*10^{-8}$	$6.2*10^{-8}$
Hydrogen	80	180	80	180	130	8.36	$1.8*10^{-7}$	$2.5*10^{-7}$

possible reference electrode drift, no consistent results could be obtained. The only possible way to estimate the crack growth rate was to use time-to-failure information from the loading diagram (Table 5.4). Crack growth values were measured from SEM micrographs on four locations per sample. Average crack growth rate was calculated using average crack increment. Maximum crack growth rate, representative of the worst-case scenario, was calculated from the highest crack increment found on the sample.

The maximum crack growth calculated for the oxygen sample was about two times faster than for the hydrogen samples. The crack growth rate for the sample tested in nitrogen was lower by an order of magnitude.

No significant change, either in fracture surface morphology or in calculated crack growth rates, was encountered in samples with applied potential compared to basic hydrogen conditions.

5.4 Conclusions

1 Electrochemical measurements indicate that addition of oxygen or nitrogen increases corrosion potential and decreases stability of the oxide layer.

2 XPS analyses and depth profile measurements of protective layers show almost similar layer composition for all environments. The oxide layer formed under oxygen conditions is thicker than the one formed under standard PWR conditions.

3 Based on the XPS results, the corrosion potential does not have a significant influence on the composition of the oxide layer. However, the corrosion potential has an influence on the oxide layer thickness.

4 SSRTs show high ductility and no presence of SCC on the sample tested in oxygen. The sample tested in pure PWR conditions shows high degree of hardening, decrease in plastic properties and extensive presence of transgranular SCC on the fracture surface. Properties of the sample tested in nitrogen are similar to the pure PWR test, with higher ductility and smaller degree of SCC on the fracture surface.

5 The highest crack growth rate calculated from ligaments of the CRBs was found on the sample tested in oxygen conditions. No significant change was found in samples with applied potential compared to basic hydrogen conditions.

6 The summary of both SSRT and CRB test results would indicate that under oxygen conditions the SCC crack initiates slowly, but once it is initiated, it propagates much faster than under hydrogen conditions.

5.5 Acknowledgments

The authors would like to thank Mr L. Eysermanns for preparation of autoclave tests and Dr A. Franquet from VUB Brussells for XPS analyses.

5.6 References

1. R.-W. Bosch *et al.*, *Corrosion* **45**, 162 (2003).
2. B. Beverskog *et al.*, *Corrosion* **55**, 1077 (1999).
3. E. Schuster *et al.*, *J. Nucl. Mater.* **152**, 1 (1988).
4. J. Robertson, *Corros. Sci.* **32**, 443 (1991).
5. Y. Sandler, *Corrosion* **35**, 205 (1979).
6. T. Nakayama and Y. Oshida, *Corrosion* **24**, 336 (1968).
7. M. Cunha Belo *et al.*, *Corros. Sci.* **40**, 447 (1998).
8. M. Scibetta, Fracture toughness evaluation of circumferentially cracked round bars, *Progress Report SCK·CEN*, Belgium, BLG-716 (1996).

6

High temperature electrochemical impedance spectroscopy of metals related to light water reactor corrosion

GABOR NAGY, ZSOLT KERNER, JÁNOS BALOG and ROBERT SCHILLER,
KFKI Atomic Energy Research Institute, Hungary

6.1 Introduction

The interest towards more severe operating conditions and the lifetime extension of nuclear reactors has induced theoretical and experimental investigations. Along these lines, we developed an experimental programme in the field of corrosion problems related to conditions which prevail in a VVER (Voda-Vodyanoi Energetichesky Reaktor) primary circuit; our interest has been focused on high temperature oxidation kinetics of zirconium (Zr) alloys and high temperature reference electrode testing. In this contribution the results of in-situ temperature-dependent electrochemical impedance spectroscopy (EIS) are discussed in detail.

Impedance spectroscopy can be applied *ex situ* on pre-oxidised samples and *in situ* during oxidation. There is no general consensus in the literature about the interpretation of impedance spectra of Zr alloys, indicating that the analysis of results in terms of equivalent circuits is not straightforward. Among others, Wikmark *et al.* [1], Göhr *et al.* [2], Bardwell and McKubre [3], Gebhardt [4], Albinet *et al.* [5] and Schefold *et al.* [6] contributed to the better understanding of EIS results. By measuring oxidised Zr-1%Nb *ex situ* in potassium sulphate solution, Krysa [7] found a complicated equivalent circuit to be valid. Vermoyal *et al.* compared oxide on Zircaloy-4 and Zr-1%Nb by impedance spectroscopy in gaseous atmospheres [8]. Photoelectrochemical investigations of Macdonald *et al.* [9] and Lee *et al.* [10] show that ZrO_2 and the passive film on Zircaloy-4 are n-type semiconductors.

In our previous work we used a large number of techniques for the study of the corrosion processes on zirconium alloys and the characterisation of the protective oxide layer of these materials [11]. We found the oxide layer to be non-homogeneous. Below the water/oxide interface the zirconium content is almost constant in the ZrO_2 phase. The thickness of this phase increases with oxidation time. Further towards the metal the oxygen content decreases linearly as a function of depth within a non-stoichiometric ZrO_x phase. The thickness of this phase also increases with oxidation time. A considerable amount of oxygen is dissolved in the alloy in the range of 10 μm below the oxide layer. The oxide layer was also investigated by photo-electrochemical measurements [12]. This method also provides quantities which are characteristic of the oxide layer; we could show that the oxide phase is an n-type semiconductor. With series of high temperature electro-chemical impedance spectroscopic measurements we hoped to follow the oxidation of Zr-1%Nb *in situ* [13]. In the present contribution we are going to discuss the temperature dependence of the EIS spectra for both the oxidised and the non-oxidised Zr-1%Nb in detail.

The importance of redox potential measurements in high temperature, high pressure aqueous systems, such as boiling water reactors (BWRs) and pressurised water reactors (PWRs), is well known. The principles of the theoretical background in this field are well known; the main problem is the reliability of the electrodes for long-term usage in such an aggressive environment [14]. There are several types of internal reference electrodes that have potential application in light water reactor (LWR) systems. Internal Ag/AgCl electrodes have already been tested successfully [15]. An Ag/ Ag_2SO_4 electrode has also been developed [16]. The yttrium-stabilised zirconium oxide electrodes have also been proved to be applicable [17]. Other types of metal/metal oxide electrodes have also been tested [18]. Reference electrodes containing polymer electrolytes are also promising [19]. Another group of internal reference electrodes consists of the hydrogen electrodes. The best known example of this type is the palladium hydride electrode, sometimes in alloy forms [20]. Other metal electrodes have also been used [21]. New types of electrodes are also under development [22].

Considering the state-of-the-art in this field the main objective of the LIRES project, supported by the EC within the EURATOM FP5, was to develop such a reference electrode which is robust enough to be used inside an LWR. Our contribution to the project was to test three types of reference electrodes: (i) external Ag/AgCl reference electrode from NRI Rez, Plc; (ii) Pt-Ir alloy hydrogen electrode from Studsvik Nuclear AB; (iii) Pd-hydride reference electrode [Pd(Pt) double polarised active electrode] from VTT Technical Research Centre of Finland [23]. In this contribution we consider the temperature dependencies of the EIS spectra in detail.

In both research areas we came across the problem of understanding the temperature dependencies of impedance spectra. By analysing the results we found tendencies which contradicted the expectations based on simple theoretical considerations and, in some cases, certain elements of the equivalent circuits have unexpected trends. The main objective of this chapter is to present and discuss these results.

6.2 Experimental set-up

The experimental set-up consisted of a PARR 4532 type 2 litre autoclave with a maximum pressure of 137 bar. The autoclave was fitted with gas inlet and outlet, manometer, safety valve with Inconel Rupture Disc, and a thermocouple. The operating temperature was monitored with a TES 1300 digital thermometer.

We tested Zr-1%Nb tubes, a Pd(Pt) active electrode pair in a Conax MHC4-062-4-L feedthrough, an external Ag/AgCl electrode with Parker M6MSC″N feedthrough, a Pt and a PtIr electrode with Parker M6MSC″N feedthrough. The Pd active electrode was polarised galvanostatically with 32 μA for producing hydrogen with a constant rate on the Pd surface. As counter electrode we used Zr-1%Nb tube or Pt wire.

We used Solartron 1286 and Electroflex EF453 potentiostats coupled to a Solartron 1250 frequency response analyser, or Autolab PGStat30 potentiostat and frequency response analyser for the electrochemical measurements. Electrochemical impedance spectroscopy was carried out in the frequency range of 64 kHz–0.1 Hz (for control experiments 0.001 Hz) with a.c. amplitude of 5 mV. The measurements were done in a two-electrode arrangement at open circuit potential [except for the Pd(Pt) double-polarised active electrode, for which the amperostatic condition of 32 μA was kept] at 290°C during testing of the electrodes, and around 20, 110 and 200°C during heat-ups.

The experiments were carried out in solutions similar to those in VVER and PWR primary circuits. Zr-1%Nb samples were measured in solutions containing boric acid with ammonium hydroxide and potassium hydroxide for VVER conditions, while the reference electrodes were tested in solutions containing boric acid with lithium hydroxide for PWR conditions, respectively. The oxygen content of the electrolyte solutions was reduced before the start of each experiment at room temperature by argon gas bubbling for at least 30 min.

6.3 Results and discussion

Under our experimental conditions, impedance spectra do not indicate complicated electrode behaviour. The spectra can usually be described in terms of a rather simple -R_s-R_p||CPE- combination in a wide frequency region of 0.1 Hz–10 kHz. Here R_s is the solution resistance, R_p is a parallel resistance, while CPE is the so-called constant phase element with two parameters, $Z_{CPE} = (1/\sigma)(i\omega)^{-\alpha}$, where σ is the CPE coefficient and α is the CPE exponent (ω is the angular frequency and i is the imaginary unit).

The spectra of non-oxidised Zr-1%Nb with native oxide on the surface (Fig. 6.1) reflect a single interfacial process, which is described by the above-

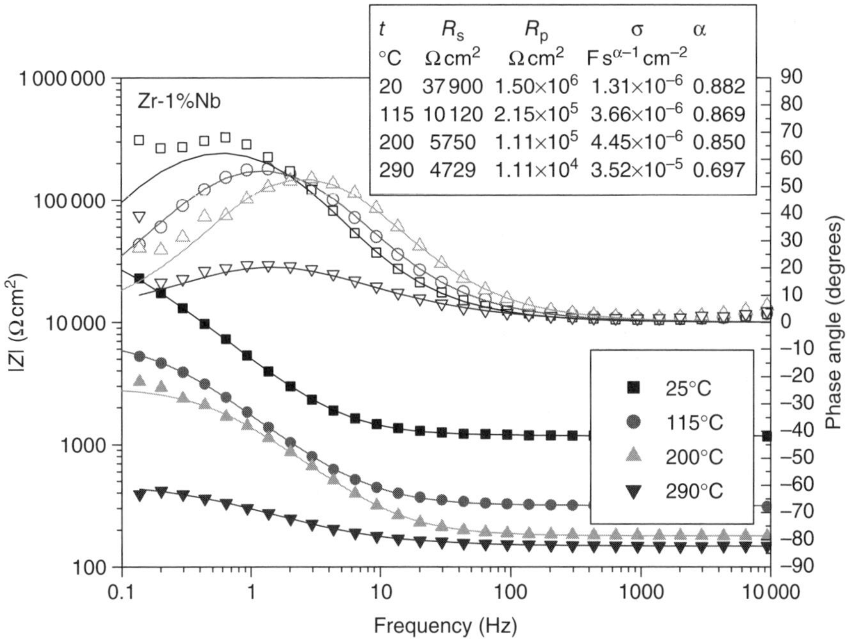

Fig. 6.1 Impedance spectra of non-oxidised Zr-1%Nb at four different temperatures. The solid lines are fitted curves to the -R_s-R_p||CPE- equivalent circuit. The fitted parameters are given in the inset.

discussed equivalent circuit. The absolute value is constant at high frequencies giving the value of solution resistance. The phase angle has a maximum in the range of 0.1–10 Hz, depending on temperature. The fitted values to the -R_s-R_p||CPE- combination are given in the inset. It was found that R_s decreases with increasing temperature, as expected for a solution resistance. R_p also decreases with increasing temperature. The CPE coefficient, σ, increases with increasing temperature, while the CPE exponent, α, decreases.

As oxidation proceeds, new features appear in the spectra (Fig. 6.2) at higher frequencies. Thus, we could attribute this part (the new peak on the phase angle diagram) to the presence of a thick oxide layer. The changes in the low frequency part may be related either to a diffusion-limited process (with semi-infinite, or transparent boundary conditions) or to a process with long characteristic time. Since our preliminary aim was to understand oxide layer properties, we analysed the results only in the 10 Hz–10 kHz frequency range by fitting the above-introduced equivalent circuit. The fitted values to the -R_s-R_p||CPE- combination are given in the inset. It is clear that the parallel resistance R_p can be attributed to the

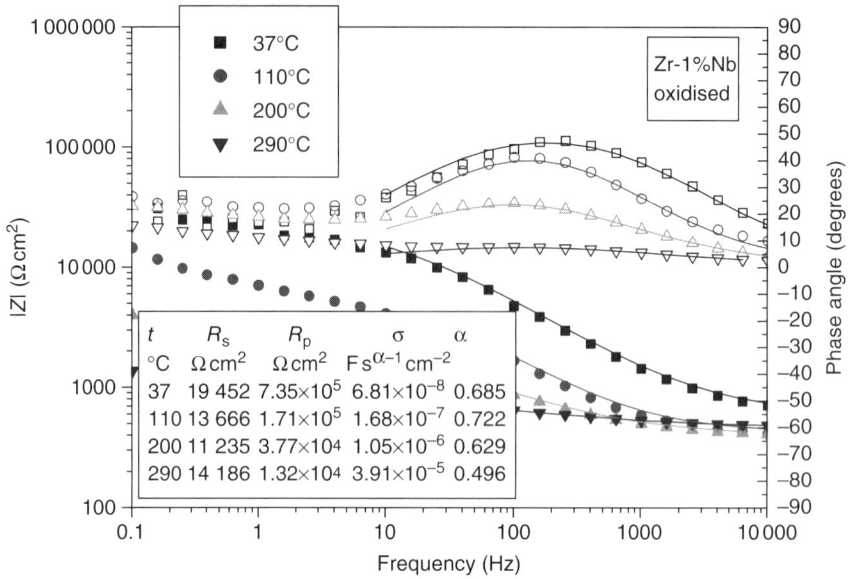

Fig. 6.2 Impedance spectra of oxidised Zr-1%Nb at four different temperatures. Filled symbols are the absolute values and open symbols are the phase angle values, respectively. The solid lines are fitted curves to the -R_s-R_p||CPE- equivalent circuit in the frequency range 10–1000 Hz. The fitted parameters are given in the inset.

electric resistance of the oxide, R_{ox}, and the CPE element is also connected to the oxide layer CPE = CPE_{ox}.

By measuring EIS in the function of oxidation time, we could conclude that the method is capable of following the oxidation process *in situ*. One can see that the CPE_{ox} coefficient, σ_{ox}, decreases, whereas the CPE_{ox} exponent, α_{ox}, is constant (Fig. 6.3), and the parallel resistance, R_{ox}, increases (Fig. 6.4). These trends are in accordance with expectations: the CPE element may reflect the capacitive behaviour of the oxide layer, thus σ_{ox} decreases with increasing layer thickness. R_{ox}, reflecting the electrical conductivity, should increase with increasing thickness. Since the R||CPE element may correspond to the charge transport through the oxide layer, we calculated a characteristic relaxation time, $\tau_{ox} = (R_{ox} \cdot \sigma_{ox})^{1/\alpha}$. Figure 6.4 shows that τ_{ox} is practically independent of oxidation time, indicating that after the first couple of days the chemical properties of the oxide layer do not change during corrosion (within our time scale).

Usually CPE_{ox} is considered to be a dispersive capacitance. In this context σ is regarded as a *pseudo*-capacity, although its dimension differs from farad, and α is the measure of the capacitive nature of the element: if $\alpha = 1$ the element is an ideal capacitor, if $\alpha = 0$ it behaves as a frequency-independent ohmic resistor, whereas if $\alpha = -1$ it behaves as an inductance.

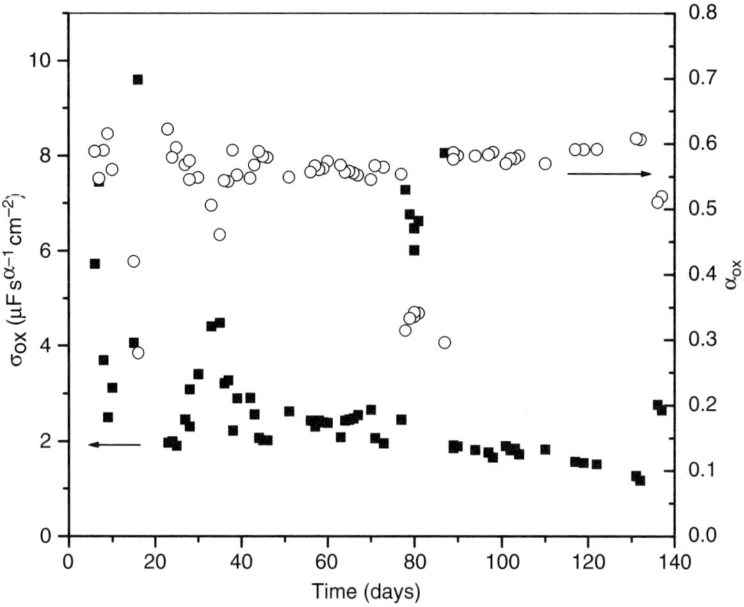

Fig. 6.3 Time dependence of the CPE parameters obtained by fitting the impedance spectra of Zr-1%Nb measured at 290°C with the R_s-$CPE_{ox}\|R_{ox}$ equivalent circuit. CPE coefficient, σ_{ox} (filled squares); CPE exponent, α_{ox} (open circles).

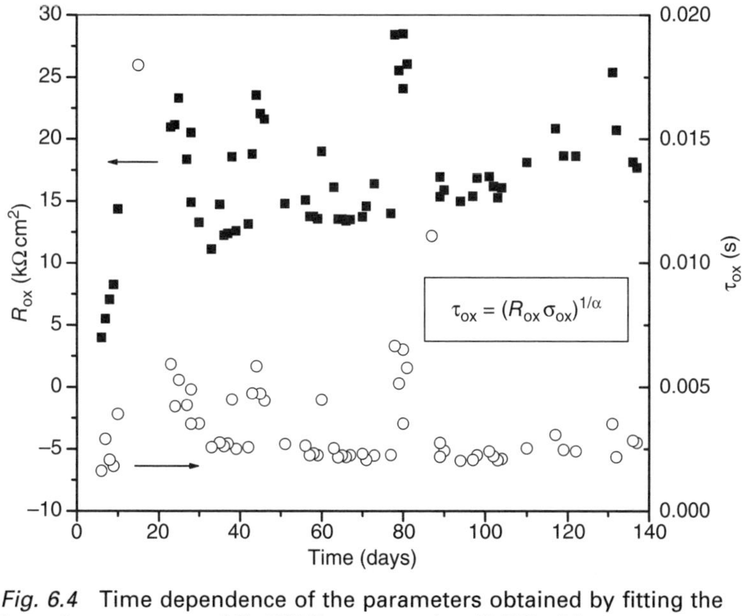

Fig. 6.4 Time dependence of the parameters obtained by fitting the impedance spectra of Zr-1%Nb measured at 290°C with the R_s-$CPE_{ox}\|R_{ox}$ equivalent circuit. R_{ox} (filled squares); characteristic time, τ_{ox} (open circles).

(Note that there are known cases when Z_{CPE} cannot be treated this way; e.g. the Warburg impedance, which is connected with diffusion, is a CPE element with $\alpha = 0.5$.) However, the temperature dependencies of these fitted parameters, as shown in Fig. 6.5, contravene any explanation that would rest on a capacitor model. The capacity of a condenser of any geometry is proportional to the relative permittivity of the dielectric material, ε. Since ε usually decreases with increasing temperature (with phase transitions representing obvious but rare exceptions), increasing temperature would make the capacity decrease, i.e., within the framework of the dispersive capacity model σ is expected to decrease as T increases. Our observations contradict this expectation.

Hence it seems reasonable to make use of a model of dispersive resistance instead of dispersive capacitance. As Macdonald [24] pointed out, the theory of continuous time random walk (CTRW) by Scher and Lax [25] can be applied to the present problem of charge transport in solid disor-

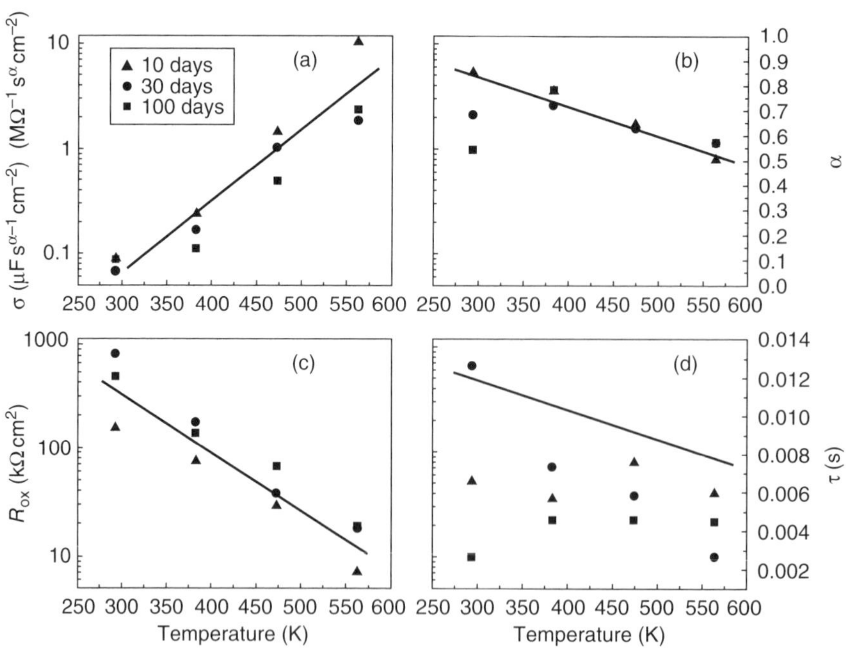

Fig. 6.5 Temperature dependence of the parameters obtained by fitting the impedance spectra of oxidised Zr-1%Nb with the R_s-CPE$\|R_{ox}$ equivalent circuit in the 10 Hz–10 kHz frequency range. (a) CPE coefficient, σ; (b) CPE exponent, α; (c) R_{ox}; (d) characteristic time, τ_{ox} (see text). Filled triangles: sample oxidised for 10 days; filled circles: sample oxidised for 30 days; filled squares: sample oxidised for 100 days. The solid lines are for guiding the eye.

dered media. Compte [26] and Bisquert and Compte [27] analysed a number of models and theories of anomalous diffusion in terms of fractional calculus, with a view to electrochemical impedance spectra. They proved that the mechanism and conditions of diffusion define the spectrum, predicting CPE behaviour under certain circumstances.

By applying this theory we could obtain a coherent picture [28] about the oxide layer. The high frequency time scale of the charge transport process, τ, is in the order of 10^{-2} s. The low frequency time scale, is defined as the average waiting time characteristic to d.c. charge carrier motion, $\theta \cong 1$ s. $1/R_{ox}$ can be regarded as a measure of oxide growth rate. The increase of $1/R_{ox}$ with increasing temperature can be attributed to both the increase of mobile charge density, and the decrease of atomic hopping time, θ. The parameters of CPE, τ and α, are thought to characterise the non-Faraday charge transport across the oxide. The orders of magnitude difference between θ and τ together with the negligible temperature dependencies of both α and τ show that the charge carriers, which are due to a.c. conductivity, are different from the atomic carriers which account for R_{ox}.

A similar situation to the one described above can be found for the Pt and Pt-Ir electrodes. As shown in Fig. 6.6, the impedance spectra have similar features. They can adequately be described with the -R_s-R_p||CPE-combination (the fitted values are given in the inset). Again, it is found that R_s decreases with increasing temperature, as expected for a solution resistance. R_p also decreases with increasing temperature. The CPE coefficient, σ, increases with increasing temperature, while the CPE exponent, α, decreases.

These findings are similar to the results for the non-oxidised Zr-1%Nb. The temperature dependence of R_s and R_p may be understood based on conventional arguments but the observation of increasing σ with increasing temperature is not consistent with the findings of negative temperature coefficients of the double layer capacitance at the pzc (see Ref. 29 and references therein). Thus, we are inclined to conclude that the conventional assignment of impedance spectra to double layer properties may not be used. We think that under our experimental conditions there may be a thin oxide layer on the platinum surface and that this give rise to the same temperature dependence of the impedance spectra as in the case of the non-oxidised Zr-1%Nb with native oxide layer. The presence of oxide layer on platinum was demonstrated by Yajima [30].

Impedance spectra of Pd (Fig. 6.7) can be described phenomenologically by making use of the same equivalent circuit. But in this case the CPE coefficient, σ, has opposite temperature dependence: it decreases with increasing temperature. This indicates that the experimental data now reflect double layer phenomena. Since Pd is cathodically polarised relative to a Pt counterelectrode [this is why it is called Pd(Pt) active electrode],

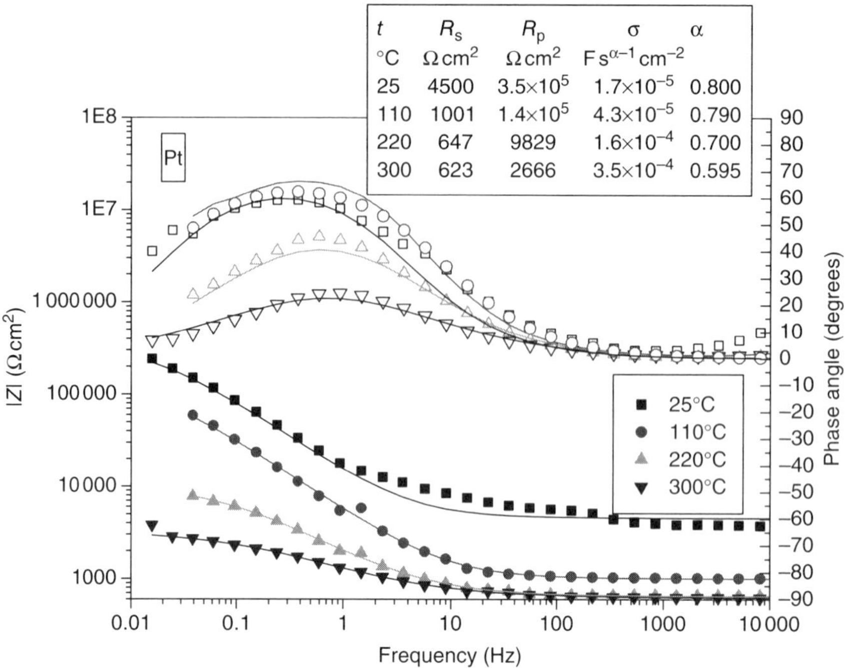

Fig. 6.6 Impedance spectra of Pt electrode at four different temperatures. Filled symbols are the absolute values and open symbols are the phase angle values, respectively. The solid lines are fitted curves to the -R_s-R_p||CPE- equivalent circuit. The fitted parameters are given in the inset.

there is hydrogen evolution on the electrode surface and the Pd is obviously metallic. Thus, we may expect the double layer capacitance to appear in the impedance spectra, an element which is supposed to have the same temperature dependence as we observed experimentally for σ.

Finally, for the sake of completeness, we present the impedance spectra of the Ag/AgCl electrode at room temperature and at 300°C in Fig. 6.8. The results indicate that the a.c. response of this electrode is determined mostly by its geometrical parameters defined by its design. The large resistance is due to the distilled water layer and a ceramic frit which separate the Ag/AgCl chamber from the environment. Clearly, this resistance decreases with increasing temperature. The capacitance value is tentatively identified with the system's stray capacitance.

6.4 Conclusions

Some problems related to corrosion under VVER primary circuit conditions are the subject of the present chapter. We had a two-fold aim: (i) to

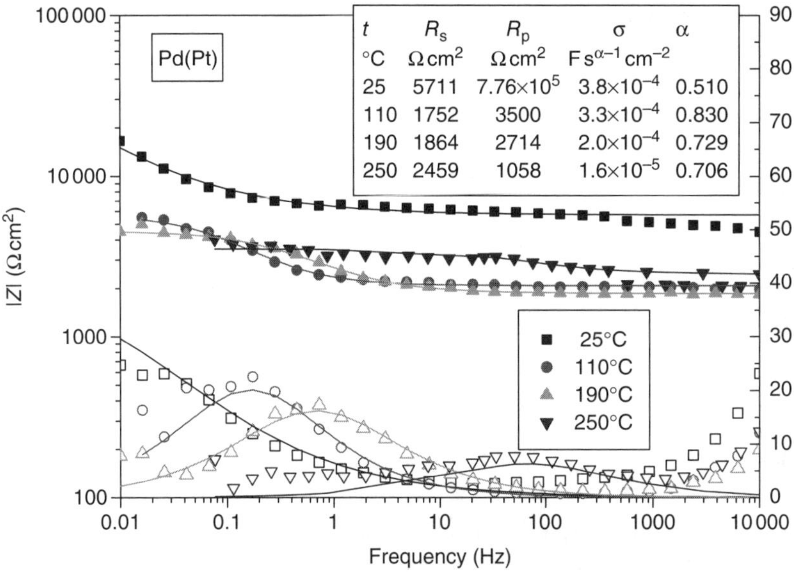

Fig. 6.7 Impedance spectra of Pd(Pt) active electrode at four different temperatures. Filled symbols are the absolute values and open symbols are the phase angle values, respectively. The solid lines are fitted curves to the -R_s-R_p||CPE- equivalent circuit. The fitted parameters are given in the inset.

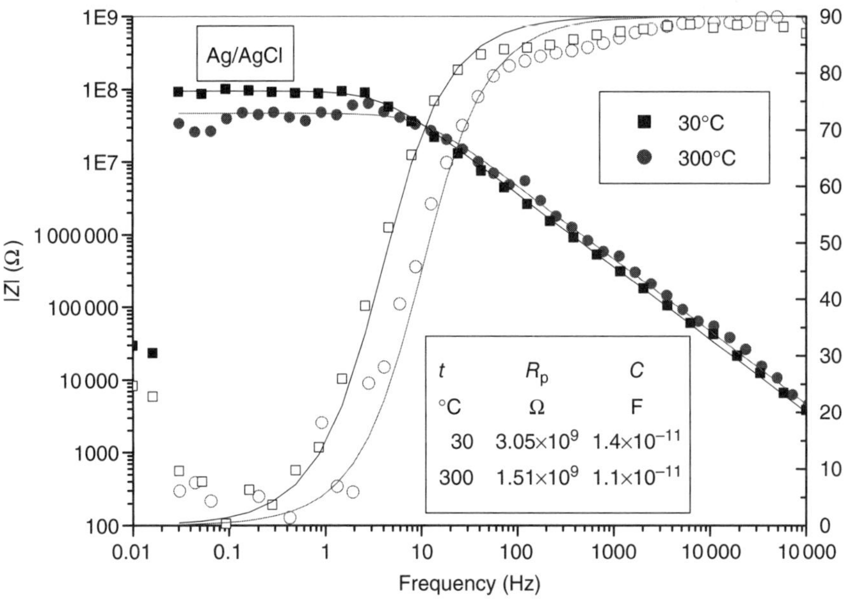

Fig. 6.8 Impedance spectra of the Ag/AgCl electrode at two different temperatures. Filled symbols are the absolute values and open symbols are the phase angle values, respectively. The solid lines are fitted curves to the -R_p||C- equivalent circuit. The fitted parameters are given in the inset.

apply electrochemical methods in order to characterise the hydrothermally produced oxides of the cladding material (Zr-1%Nb) of nuclear fuel elements used in Russian-made type VVER power reactors, and (ii) to investigate a number of possible reference electrodes with a view to high temperature applications.

Impedance spectroscopy proved to be a very useful technique to characterise these systems. From temperature-dependent measurements we could draw some conclusions about the nature of the rate-determining interfacial process by analysing the trends of the equivalent circuit parameters. We found that different double layer properties may manifest themselves in similar impedance spectra. We were able to show that:

1 The a.c. response of the oxidised Zr-1%Nb is due to the charge transfer process in the oxide phase.
2 Probably the same effect is reflected in the impedance spectra of the non-oxidised Zr-1%Nb, the Pt and the Pt-Ir electrodes due to the presence of a native oxide layer on the surface.
3 A real double layer contribution can be found for the cathodically polarised Pd electrode.
4 The impedance of the Ag/AgCl electrode is determined by its design parameters.

We believe that our results show the benefits of electrochemical impedance spectroscopy, which is a very powerful technique for the understanding of electrochemical systems. The temperature-dependent measurements helped us to identify the most important electrochemical processes for a wide range of systems of practical importance.

6.5 Acknowledgments

The financial support of the European Union within the 5th Framework Programme for the LIRES project, under contract no. FIKS-CT-2000-00012, is gratefully acknowledged. G.N. is indebted to the Alexander von Humboldt Foundation for a return research fellowship. This work was also supported by the National Science Research Fund, Hungary (OTKA) under Contract Nos. T049202 and F049839.

6.6 References

1. G. Wikmark, P. Rudling, B. Lehtinen, B. Hutchinson, A. Oscarsson and E. Ahlberg, *Zirconium in the Nuclear Industry: Eleventh Int. Symp., ASTM STP 1295*, ed. E. R. Bradley and G. P. Sabol, ASTM, 1996, pp. 55–73.
2. H. Göhr, J. Schaller, H. Ruhmann and F. Garzarolli, *Zirconium in the Nuclear Industry: Eleventh Int. Symp., ASTM STP 1295*, ed. E. R. Bradley and G. P. Sabol, ASTM, 1996, pp. 181–202.

3. J. A. Bardwell and M. C. H. McKubre, *Electrochim. Acta*, 1991, **36**, 647–653.
4. O. Gebhardt, *J. Nucl. Mater.*, 1993, **203**, 17–26.
5. B. Albinet, B. Sala, A. Frichet and H. Takenouti, *7th Int. Symp. Electrochemical Methods in Corrosion Research*, EMCR2000, Budapest, Hungary, May 28–June 1, 2000, CD rom no. 086.
6. J. Schefold, D. Lincot, A. Ambard, E. Moleiro and O. Kerrec, *7th Int. Symp. Electrochemical Methods in Corrosion Research*, EMCR2000, Budapest, Hungary, May 28–June 1, 2000, CD rom no. 058.
7. J. Krysa, *ACH Models in Chem.*, 2000, **137**, 249–268.
8. J. J. Vermoyal, A. Frichet, L. Dessemond, A. Hammou, *Electrochim. Acta*, 1999, **45**, 1039; *Electrochim. Acta*, 2002, **47**, 2679.
9. A. Goossens, M. Vazqeuz and D. D. MacDonald, *Electrochim. Acta*, 1996, **41**, 47.
10. S. J. Lee, E. A. Cho, S. J. Ahn and H. S. Kwon, *Electrochim. Acta*, 2001, **46**, 2605–2611.
11. G. Nagy, G. Battistig, A. Csordás-Tóth, Zs. Kerner and T. Pajkossy, *J. Nucl. Mater.*, 2001, **297**, 62–68.
12. J. Balog, G. Nagy and R. Schiller, *Radiat. Phys. Chem.*, 2003, **67**, 321–323.
13. G. Nagy, Z. Kerner and T. Pajkossy, *J. Nucl. Mater.*, 2002, **300**(2–3), 230–236.
14. P. Bennett, 'Recent in-core ECP measurements in the Halden Reactor', Halden IASCC, Review meeting, November 2002.
15. M. E. Indig and J. L. Nelson, *Corrosion*, 1991, **47**(3), 202–209.
16. S. Lee, R. W. Staehle, *Mater. Corr. – Werkstoffe Korr.*, 1997, **48**(12), 807–812.
17. M. Navas, M. D. Gomez Briceno, *Nucl. Engng Design*, 1997, **168**(1–3), 183–189.
18. L. B. Kriksunov and D. D. Macdonald, *J. Electrochem. Soc.*, 1994, **141**(11), 3002–3005.
19. P. Jayawera, 'Stable solid state reference electrode for high temperature water chemistry', *Corrosion 95*, paper 445, NACE International, Houston, Texas.
20. V. A. Marichev, *Corrosion*, 1996, **52**(1).
21. J. A. Sampedro, N. Rosas, B. Valdez, *Corros. Rev.*, 1999, **17**(3–4), 253–262.
22. R.-W. Bosch, W. F. Bogaerts, J. H. Zheng, *Corrosion*, 2003, **59**(2), 162–171.
23. R.-W. Bosch, G. Nagy, D. Feron, M. Navas, W. Bogaerts, D. Kárník, R. Kilian, A. Molander, K. Mäkelä, 'Actual status of the LIRES-project: Development of Light Water Reactor Reference Electrodes', *Proceedings of EUROCORR 2003*, Budapest, Hungary, No. 128, 2003.
24. J. R. Macdonald, ed., *Impedance Spectroscopy, Emphasizing Solid Materials and Systems*, Wiley, New York, 1987.
25. H. Scher and M. Lax, *Phys. Rev. B*, 1973 **7**, 4491.
26. A. Compte, *Phys. Rev. E*, 1996, **53**, 4191.
27. J. Bisquert and A. Compte, *J. Electroanalyt. Chem.*, 2001, **499**, 112.
28. G. Nagy and R. Schiller, *Phys. Chem. Chem. Phys.*, 2002, **4**, 791.
29. A. Popov, N. Dimitrov, R. Naneva and T. Vitanov, *J. Electroanalyt. Chem.*, 1994, **376**, 97.
30. T. Yajima, *Surf. Coat. Technol.*, 1999, **112**, 80.

7

High temperature corrosion of Zircaloy-4 followed by *in-situ* impedance spectroscopy and chronoamperometry. Effect of an anodic polarisation

M. TUPIN, C. BATAILLON, J. P. GOZLAN and P. BOSSIS, CEA – Saclay, France

7.1 Introduction

The study of the corrosion behaviour of several cladding tubes under pressurized water reactor (PWR) conditions is performed by analysing the oxide layer after irradiation cycles. There is at present no tool to follow *in situ* the oxidation rate of the zirconium alloy. Moreover, the corrosion process depends on many parameters: neutron flux, hydrodynamic conditions, water vapour fraction, internal temperature, external pressure.

Because of the difficulties of studying in-reactor corrosion processes, studies were first aimed at the out-of-pile oxidation rate in order to get a better understanding of the corrosion mechanism and to find the rate-limiting step in these conditions. Since the beginning of the 1990s, many surveys have been carried out in this way [1–4]. At low water vapour pressure, the thermogravimetry is an efficient technique to follow continuously the weight gain [5]. Concerning the corrosion in primary medium, the oxidation rate is usually studied by measuring the weight of the sample at different times, which is clearly a discontinuous method. That is the reason why other techniques have been tested in order to acquire a signal related to the corrosion rate in real time inside a static autoclave. Impedance spectroscopy is one of the most convenient and widely used methods to study *in situ* the oxidation process of zirconium alloy in primary medium. Many articles have been written with the aim of showing the efficiency of this technique [6, 7]. Moreover, some experimental devices (e.g., oxidation loops) are devoted to the study of the influence of hydrodynamic conditions or thermal flux on the corrosion rate. Finally, by using all these techniques, the impact of most of the determining rate parameters can be evaluated, quantified and thus dissociated from each other. As a consequence, as soon as these effects are separately estimated, we should be able to deduce the real impact of neutron flux on the corrosion process under irradiation.

In the light water reactor Zircaloy-4 has been the most used alloy, but in the last few years new cladding materials have appeared to replace Zircaloy-4. However, this study specifically deals with the Zircaloy-4 corrosion process.

Whatever the operating corrosion conditions (oxygen, steam, water), the Zircaloy-4 kinetic curve is composed of a series of sub-parabolic regimes [7, 8], which means a regime slower than a pure diffusion-controlled process, i.e., parabolic regime. As a consequence, it is generally assumed that oxygen diffusion is the rate-limiting step during the whole corrosion process [9–11]. We aim to test whether our experimental results are consistent with this assumption.

Concerning the aims in this study, the first objective was to design and construct a simple device able to be nuclearised in a hot line. Secondly we wanted to know if the polarisation had an effect on the corrosion rate and if the anodic current of the polarised sample obtained by chronoamperometry was related to the current of the sample remaining at the free corrosion potential.

During this work, we have actually oxidised two samples at the same time and followed, by impedance spectroscopy, their oxidation rate. After 27 days, one of the samples was polarised in order to access the anodic current variations during the corrosion process.

For those who want to know more about the theoretical basis of electrochemical measurements, we advise reading first the technical report concerning the 'Identification of electrochemical processes by frequency response analysis' [12].

In this chapter, the operating conditions, the studied specimen and the experimental device used to acquire the impedance spectra and the anodic currents are first described by insisting on the simplicity and the advantages of this system. Secondly, the experimental results obtained with the sample at the free corrosion potential, called unpolarised or 'UP sample', are exposed and analysed from a kinetic point of view. The potential variations of the samples as well as the electrolyte resistance evolution are exhibited; moreover, the Cole–Cole transform and the method of thickness layer determination are explained. The negative imaginary impedance variations and the real ones are finally exposed in order to show the kinetic stages. The next part is devoted to the polarised sample, called the 'P sample', and especially to the anodic current acquired. Its variations are compared to the real conductance evolution of the UP sample. The results of the ΦE test are also presented and the typical impedance spectrum obtained with the P sample is analysed and compared to that of the UP sample. Finally the oxide layer micrographies of the samples are exhibited in order to show some differences in their thicknesses or their morphologies.

7.2 Experiment

7.2.1 Operating conditions and material

Two samples of stress-relieved Zircaloy-4 have been oxidised for a period of approximately 103 days. The temperature was set at 355°C, the total pressure in the autoclave was 18.7 MPa. The partial pressure of water was equal to 17.5 MPa, that of hydrogen 0.06 MPa and the pressure of inert gas was approximately 1 MPa at 355°C. Moreover, the water composition was close to the primary media found in a PWR reactor, 1000 ppm of boron in the state of boric acid and 1 ppm of lithium as lithium hydroxide.

The chemical composition of the studied stress-relieved Zircaloy-4 is indicated in Table 7.1.

7.2.2 Device improvement

Because of the high pressure, it is necessary to use an autoclave (in 316 L) for studying the cladding corrosion. The main technological difficulty for such an experimental system is to avoid gas leakage and so to maintain the water tightness of the autoclave.

The first system used was composed of three electrodes per electrical feed-through; however, this device showed inconveniences. First of all, the current wires were welded to the cladding tubes and these connections were not very tough because of the localised corrosion at these spot welds. Moreover, the electric insulation of the different wires was ensured by many stabilised zirconia ceramic seals, in which the wires had to be stringed. This step was particularly long and laborious and its feasibility would not have been possible in a hot cell. That is the reason why a simplified device has been conceptualized. This new system requires only one wire for each electrical feed-through and just a long alumina tube for the electric insulation between the working electrode and the counter-electrode, which is in this case the autoclave. The pseudo-reference electrode is a cylindrical grid of platinum with a high surface using one electrical feed-through as well.

The water tightness of the electrical feed-throughs is ensured by a PTFE seal element. Because of the fact that the PTFE creeps beyond 200°C, cooling systems have been installed under these electrical feed-throughs.

Table 7.1 Chemical composition of stress-relieved Zircaloy-4 (wt %)

Cr (wt%)	Fe (%)	O (%)	Sn (%)	Zr
0.105	0.21	0.132	1.27	Bal.

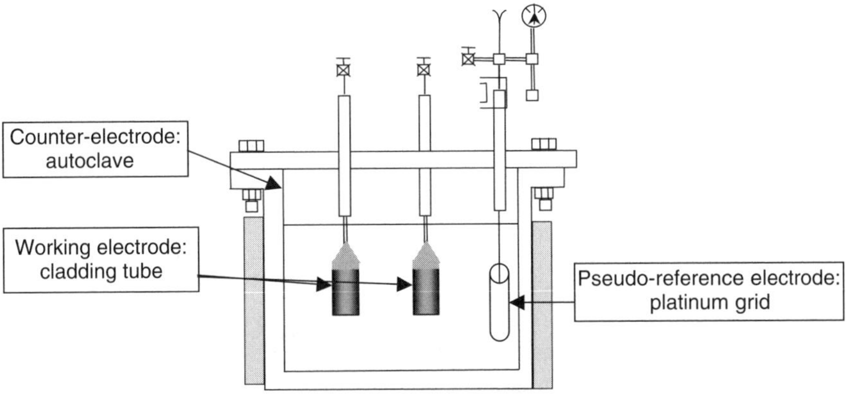

Fig. 7.1 Schematic drawing of the electrochemical device.

Concerning the electric connection between the sample and the wire, the cladding tube is welded on a pressure cap used in the nuclear industry for the fuel rod. This element is also welded at three different points with a laser to the connection wire. In this case, the device is much tougher than the last one and clearly easier to perform, especially in hot line.

The advantage of this device is its simplicity and this working electrode system is very close to the fuel rod. As illustrated in Fig. 7.1, two samples can be studied with this system.

7.2.3 Experimental measurements

The corrosion of these two samples was first followed by impedance spectroscopy at the free corrosion potential. The impedance spectra were obtained with GAMRY software (3.2 version) and the open circuit potential is measured before acquiring these spectra. After 27 days, the potential of one of the samples was increased from –270 to 0 mV versus the reference electrode potential, while the second one stayed at the free corrosion potential. These two samples were called P for the polarised sample and UP for the non-polarised sample. The anodic current of the P sample was directly measured at the set potential by chronoamperometry, although the corrosion rate was still followed by electrochemical impedance spectroscopy (EIS) for the UP sample. Moreover, few potential drops were performed on the P sample from 0 to 100 mV versus the platinum grid potential at different oxidation times, in order to determine whether the assumption of the rate-limiting step is justified [14, 15]. Finally, impedance spectra were acquired under polarisation for comparing them to those obtained at the open circuit potential.

Furthermore, let us emphasise that the polarisation needs to be set with regard to the pseudo-reference electrode. Thus this one is no longer available

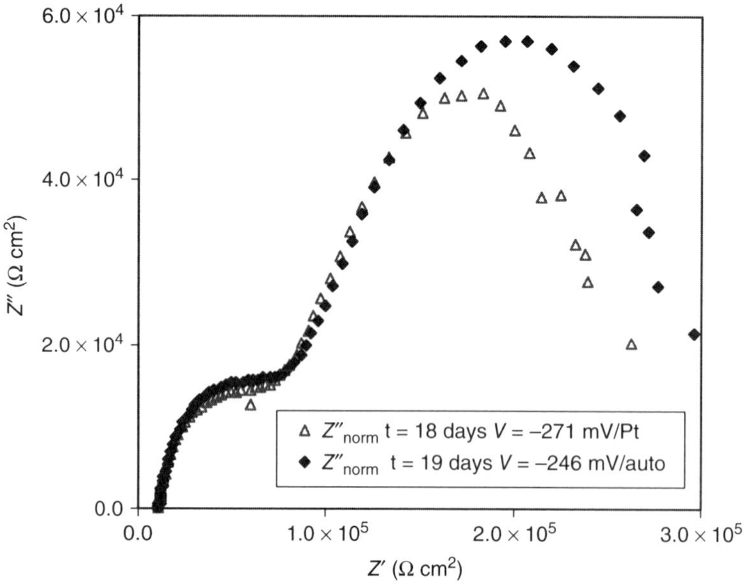

Fig. 7.2 Comparison of the spectra obtained with a two- or three-electrode acquisition system.

for acquiring the impedance spectra at the free corrosion potential for the UP sample. As a consequence, the autoclave is used as counter-electrode and reference electrode as well. As shown in Fig. 7.2, the spectra obtained with a two- or three-electrode acquisition system are nearly superimposed except at very low frequencies. The switch between the curves is actually the result of the oxidation time difference or can be due to the impedance of the autoclave can.

Z' is the real part of the impedance and Z'' the negative imaginary part.

7.3 Experimental results and interpretation

7.3.1 Electrochemical measurements

The results obtained with the UP sample will be the first to be described and analysed.

UP sample

Electrode potential variations

After about 3 oxidation days, the two working electrodes had the same potential. Concerning the counter-electrode, the autoclave potential was

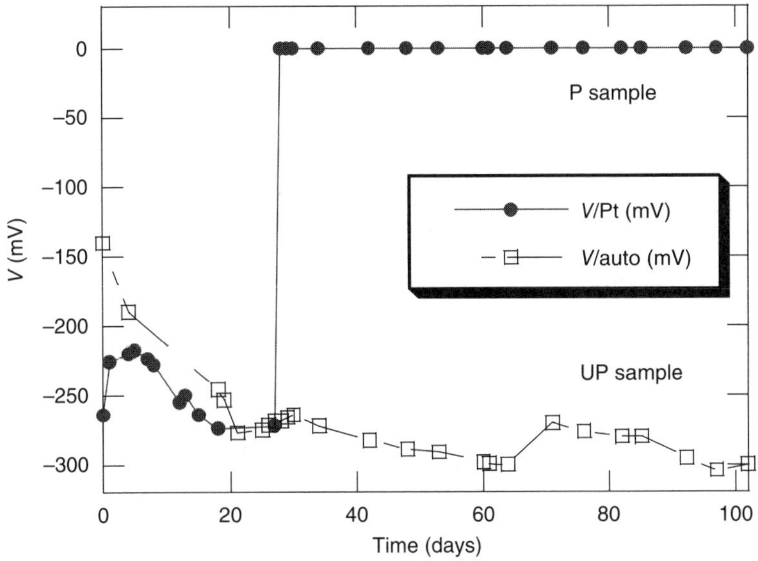

Fig. 7.3 Potential variation as a function of oxidation time. P, polarised; UP, unpolarised.

equal to that of the reference electrode after 20 days, as indicated by Fig. 7.3; the potential variations of the P sample are also indicated. Taking into account the plots for the P and UP samples, we can conclude that the difference between the free corrosion potential of the UP sample and that of the P sample is fairly constant during the corrosion process (after 27 days).

Another relevant parameter usually evaluated by impedance spectroscopy is the electrolyte resistance.

Electrolyte resistance

The electrolyte resistance is determined by extrapolating at infinite frequency the Nyquist diagram. By using for instance a polynomial to describe the curve at high frequencies, the intersection of this fitted polynomial with the abscissa corresponds to the electrolyte resistance. The extrapolated value obtained with the experiment exhibited in Fig. 7.4 is typically 240 Ω. This spectrum was performed after 90 days of oxidation.

Figure 7.5 shows the electrolyte resistance variations during the corrosion process. This parameter clearly increases with the oxidation time. This decrease of the solution conductivity may be due to the lithium or proton incorporation into the passive oxide layer of the autoclave can, whose immersed surface is around 1200 cm^2. The electrolyte resistance variations

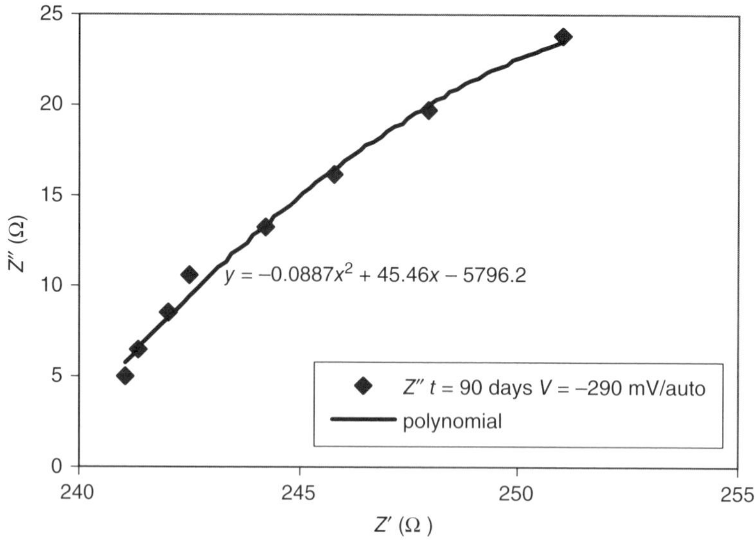

Fig. 7.4 Nyquist diagram spectrum extrapolation at infinite frequency.

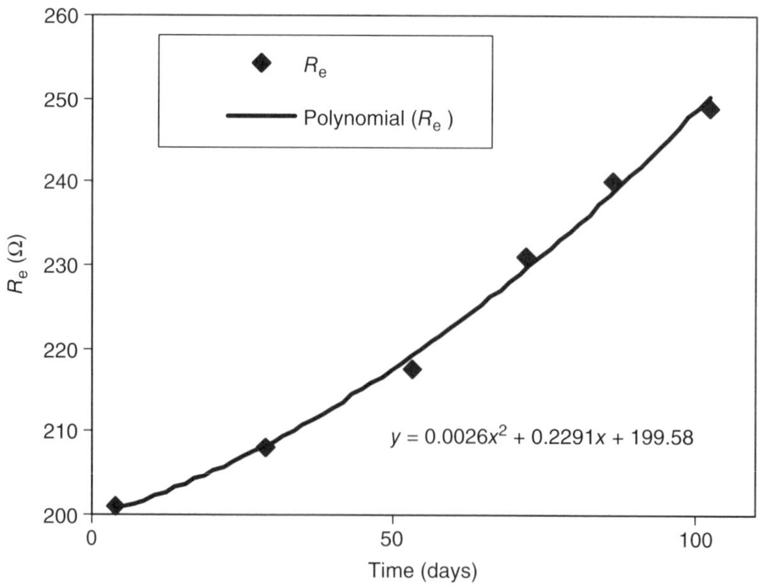

Fig. 7.5 Electrolyte resistance (R_e) variations versus time.

do not significantly modify the potential imposed on the polarised sample. We can thus consider that they are acceptable during the oxidation time.

The evaluation accuracy of the electrolyte resistance is very important because this parameter will be used for the Nyquist to Cole–Cole diagram transformation.

Cole–Cole diagram and infinite frequency capacitance

This transformation consists of defining a complex capacitance from the impedance, as follows:

$$C^* = C' - jC'' = \frac{1}{j\omega(Z(\omega) - R_e)} = \frac{1}{j\omega((Z'(\omega) - R_e) - jZ''(\omega))} \qquad [7.1]$$

where C' is the real part of the complex capacitance, C'' the negative imaginary part and ω the pulsation. It yields the following expressions for C' and C'':

$$C' = \frac{1}{\omega} \frac{Z''}{(Z'(\omega) - R_e)^2 + Z''^2} \qquad [7.2a]$$

$$C'' = \frac{1}{\omega} \frac{(Z' - R_e)}{(Z'(\omega) - R_e)^2 + Z''^2}. \qquad [7.2b]$$

The Cole–Cole diagram indicates the variations of the negative imaginary part of the complex capacitance (C'') as the function of its real part (C'). This representation specifically reveals the capacitive properties of the studied system. As shown on Fig. 7.6, the curve obtained with the UP sample is about linear, which proves that the capacitance follows the Jonscher law [15].

The oxide layer can be considered at high frequencies as a perfect capacitor. The real capacitance extrapolated at infinite frequency is then related to the thickness by the classical formula of a plane capacitor:

$$C_\infty = \frac{\varepsilon_r \varepsilon_0 S}{X} \qquad [7.3]$$

where ε_0 and ε_r are respectively the vacuum permittivity and the dielectric constant, S the electrode area and X the oxide thickness.

For this reaction system, the dielectric constant is taken to be equal to 18 at 355°C. This parameter was evaluated in order to obtain a final oxide thickness close to that determined by scanning electron microscopy (SEM). We may observe that this value is not so far from that obtained at ambient temperature, which is 22 [16]. The immersed area of the samples was estimated at 40 cm^2.

The determination of this infinite frequency capacitance is similar to the method used for the electrolyte resistance evaluation. So it is possible to

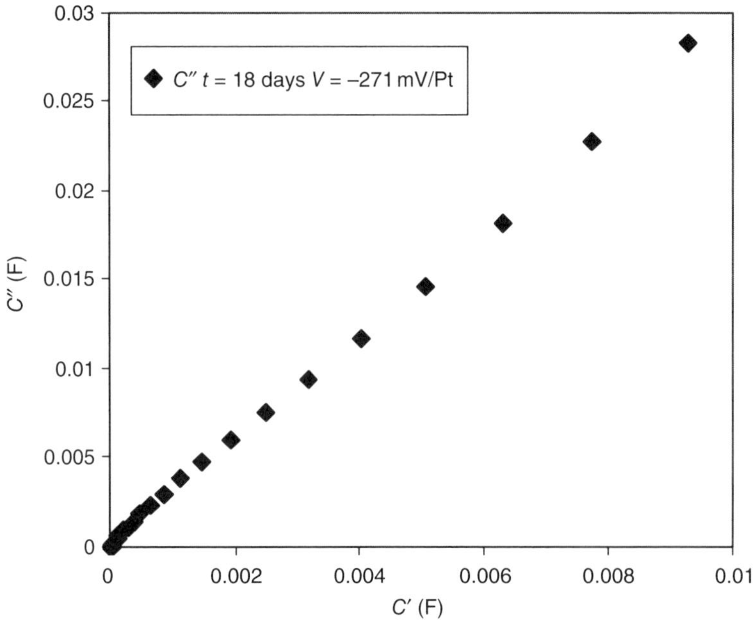

Fig. 7.6 Cole–Cole diagram for the unpolarised sample oxidised for 18 days.

determine the thickness variations with respect to the oxidation time, as shown in Fig. 7.7. According to Fig. 7.7, the thicknesses determined by impedance spectroscopy (deduced from Eq. 7.3) are consistent with the fit obtained from a typical weight gain fitting curve (cf. continuous line in Fig. 7.7). Analysing this curve, two kinetic stages can be distinguished, the pre-transition and the post-transition after about 50 days, as usually observed. The oxidation rate of the former regime monotonously decreases, while the second one takes place after the rate maximum occurring at the end of the kinetic transition. However, as shown by the error bars in Fig. 7.7, this method of determination is not very accurate. The uncertainty arises from the electrolyte resistance evaluation obtained by extrapolating the Nyquist diagram at high frequencies. Consequently, the results obtained with this method have to be compared to other results deduced from the spectra or other experimental techniques in order to verify the data consistency.

Impedance spectrum shape

Figure 7.8 indicates the typical impedance spectrum obtained with the UP sample after 27 days of oxidation, the frequency varying from 10^5 to 10^{-5} Hz. This curve shows two loops, one at high frequencies (loop 1) and the other at low frequencies (loop 2). These are not well centred and are actually

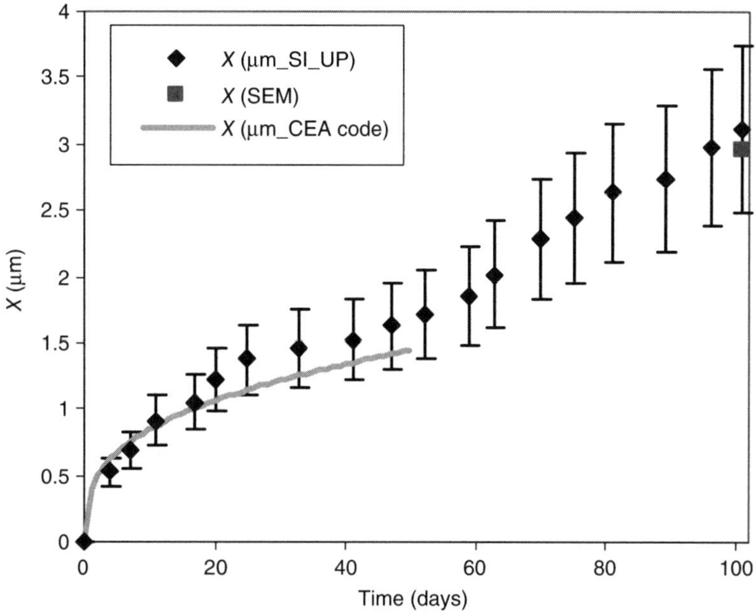

Fig. 7.7 Thickness variations determined by impedance spectroscopy (diamond plots) versus oxidation time. Comparison to the literature data (line).

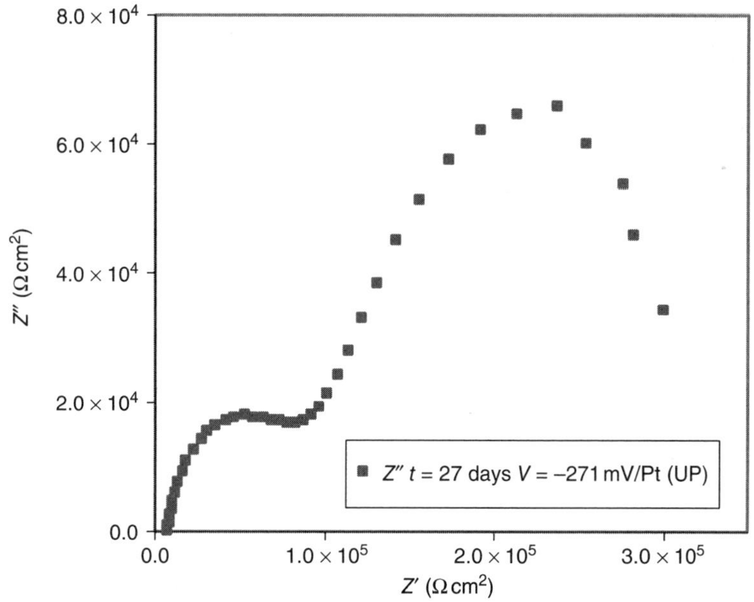

Fig. 7.8 Impedance spectrum morphology for the unpolarised (UP) sample in pre-transition.

flattened on the real axis. This whole shape is consistent with the spectra described in the literature [6, 7].

A perfect circular loop can be fitted by an $R\|C$ circuit in parallel, with a time constant equal to RC. It is also possible to describe the curve shown in Fig. 7.8 by an electric equivalent circuit but in this case it is necessary to introduce instead of a pure capacitance a complex one following the well-known empirical Jonsher law [15]:

$$C^* = C_\infty + \frac{B}{(j\omega)^{1-n}} \qquad [7.4]$$

where C_∞ is the infinite frequency capacitance, B and n two constants, n being the dispersion factor related to the slope of the Cole–Cole diagram. The second term corresponds to a CPE (constant phase element), which shows that the loop centre is not adjusted on the real axis.

Furthermore, these two loops are representative of physico-chemical processes occurring in the electrochemical system. They are defined by a kinetic constant, related to the characteristic frequency taken at the loop top. But the main difficulty with this technique is to establish the link between a particular physico-chemical process and a typical signature obtained by impedance spectroscopy. Finally it is almost always possible to represent the EIS curve by an equivalent circuit but this electric analogy is not sufficient from a kinetic point of view.

Concerning the interpretation of these loops found in the literature, Schefold *et al.* [7] considered that the high frequency loop results in the dielectric response of the total oxide layer and that the high frequency capacitance is related to the thickness of the whole layer. They also observed after the second kinetic stage the formation of a new loop at intermediate frequencies. This loop does not appear on our plots because the oxidation time is not sufficient. It is interpreted as the result of the re-passivation/degradation process of the dense layer and shows the cyclical behaviour of the Zircaloy-4 oxidation process. The work of Sala *et al.* [6] dealt more with the role of lithium and the effect of the oxygen partial pressure. Concerning the influence of these parameters on the spectra, they think that the low frequency loop is related to an interfacial step (metal oxidation, oxygen reduction or hydrogen oxidation). At present, we are still not able to confirm or reject these interpretations.

Instead of looking at the whole impedance spectrum evolution as a function of oxidation time, we have chosen to focus on the variations of a particular point, like the loop tops. Figure 7.9 presents the evolution of the negative imaginary part and that of the real one minus the electrolyte resistance taken at the high frequency loop maximum. The frequency in this case is around 2 Hz for this loop. The negative imaginary part and the real part have the same evolution during the corrosion process. We may observe in

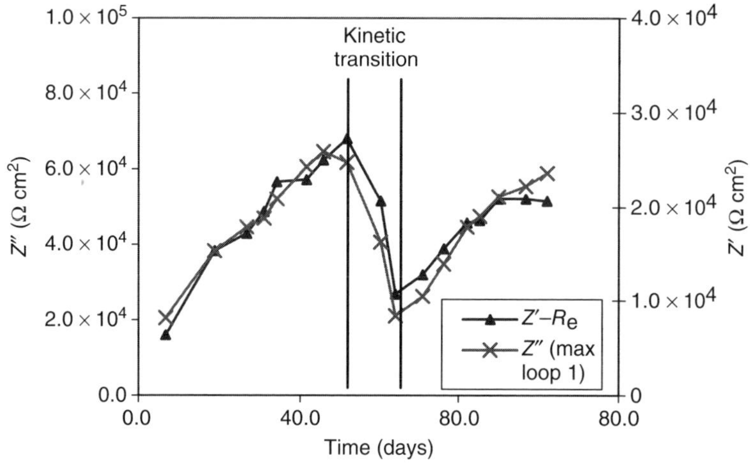

Fig. 7.9 Evolution versus time of the negative imaginary part and that of the real one taken at the high frequency loop top.

particular that after a period of increase in the pre-transition stage, the two variables considered decrease drastically at the same time. After 65 days, they increase again. Considering that the impedance real part is related to the resistance of the oxide layer in terms of electronic conductivity, its huge variations observed after 50 days are certainly the result of the oxide layer degradation observed during the kinetic transition, which involves a partial permeability of the film to the medium.

For the low frequency loop top, the conclusions would have been the same as previously.

P sample

Anodic current variations

As mentioned before, one of the two studied samples was progressively polarised from –270 mV to 0 mV/Pt, until the end of the experiment. Before describing the current variations of this sample, it is interesting to remind ourselves of the classical log(I)–V curve obtained for a metal corrosion process and to show the potential switch applied to the polarised sample (Fig. 7.10). In the figure, V_{corr} and V_{pol} correspond respectively to the corrosion potential and the anodic potential imposed on the sample.

In the activity domain (cf. Fig. 7.10), the metal is dissolved by the corrosive medium up to a critical potential, called Flade potential. In this case, the current obeys the Butler–Volmer law and varies exponentially with the potential. After the drastic drop observed beyond the Flade potential, the

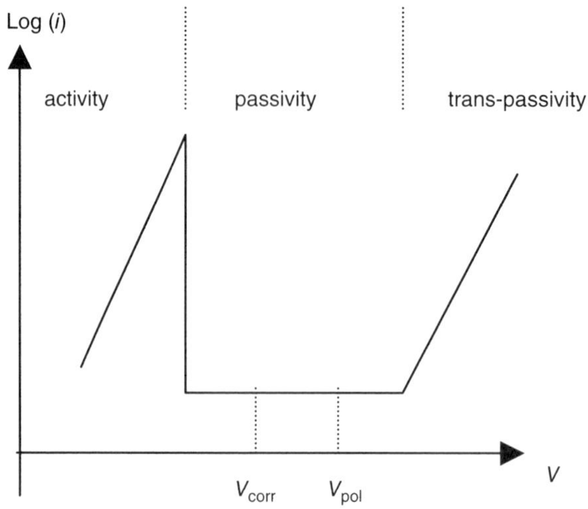

Fig. 7.10 General shape of the log (*i*)–*E* curve.

current depends slightly on the potential. The *I–V* curve actually shows a gentle slope and the metal is recovered in this case by a passivation film. This is the passivity domain. Then, beyond this plateau, other reactions can occur like an oxygen release, which produces a current increase.

Figure 7.11 describes the measured current variations during the corrosion process of the P sample, compared to the current deduced from the kinetic curve obtained at the open circuit potential and calculated from a Code developed in the CEA.

The measured current curve is slightly above that calculated from the CEA Code. The rate ratio is around 1.3 in the pre-transition stage, which first indicates that the cathodic part of the global current is negligible compared to the anodic part. The slight dependence on the potential of the anodic current proves that the free corrosion potential for Zircaloy-4 in this environment is located at the passivity plateau. Moreover, this in-situ technique of current measurement clearly allows us to observe the kinetic acceleration happening during the transition.

Moreover, the oxide thickness variations of the P sample can be calculated by integrating the anodic current curve as a function of time. The black curve on Fig. 7.12 represents this thickness evolution during the corrosion process. This one is compared to the UP sample thickness evaluated by impedance spectroscopy and to the final thicknesses determined by SEM at the end of the experiment.

According to these plots, there is no thickness difference between the two studied samples in the pre-transition stage, while a slight mismatch is observed in the post-transition stage. In this case (Fig. 7.12), the polarisation

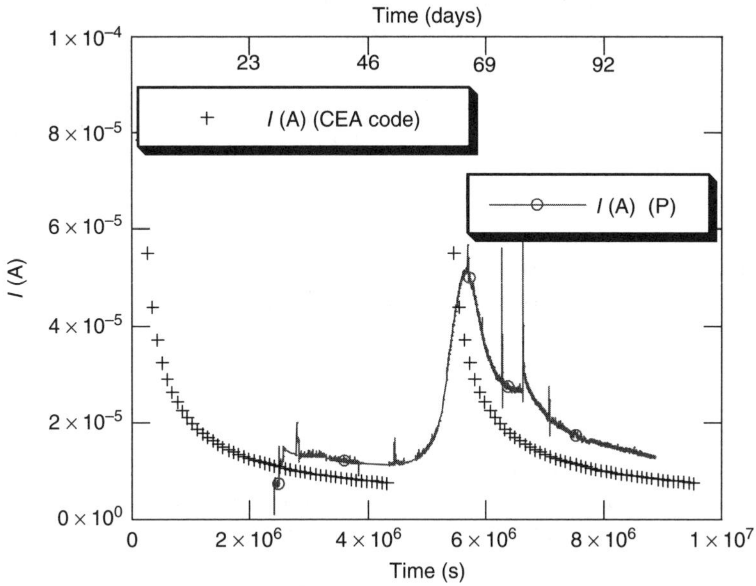

Fig. 7.11 Current evolution obtained with the polarised sample (P) and comparison to the current curve deduced from a typical kinetic curve.

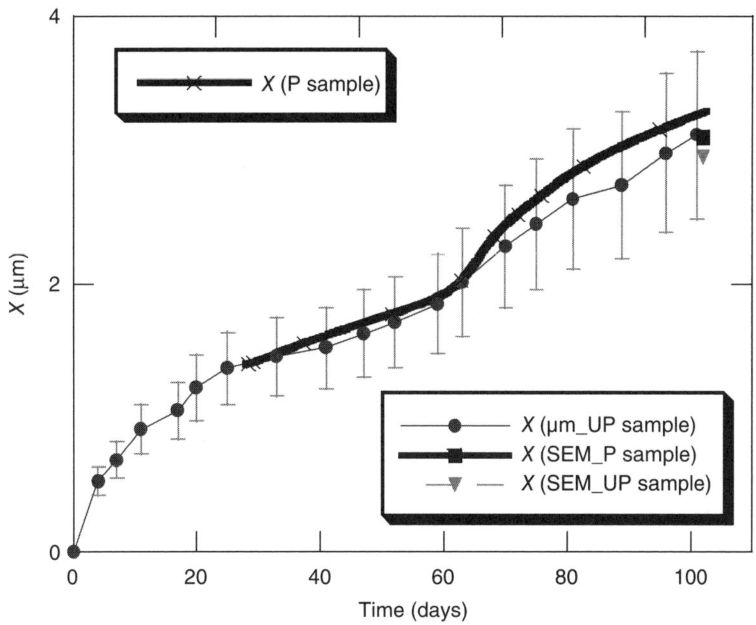

Fig. 7.12 Thickness variations for the two samples respectively obtained by integrating the anodic current curve (polarised, P sample) and by impedance spectroscopy (unpolarised, UP sample). Comparison to those obtained by scanning electron microscopy (SEM).

does not significantly affect the corrosion rate. This conclusion is confirmed by the thicknesses evaluated by SEM, which are respectively 3.1 μm for the P sample and 2.95 μm for the UP sample (cf. Section 7.3.2). However, a slope difference in pre-transition between the kinetic curves of the two samples should theoretically be found. Three reasons can be evocated to justify this discrepancy, as explained at the end of Section 7.3.2.

In order to verify that the kinetic transition occurs at the same time on the two samples, two curves have been drawn on Fig. 7.13, which describes the anodic current of the P sample and the real conductance variations of the UP sample. This real conductance is defined as follows:

$$Y = \frac{1}{(Z' - R_e)} \qquad [7.5]$$

As indicated by the curves in this Fig. 7.13, these two variables have the same evolution versus time. More specifically, the transition begins at the same time and lasts about 15 days for the two samples, which confirms once again that the kinetic stages are not significantly affected by the potential.

ΦE test

The aim of this test is to draw conclusions on the mathematical expression of the rate [5, 13, 14]. The experimental principle of this test consists of increasing suddenly the potential of the polarised sample, from 0 to 100 mV,

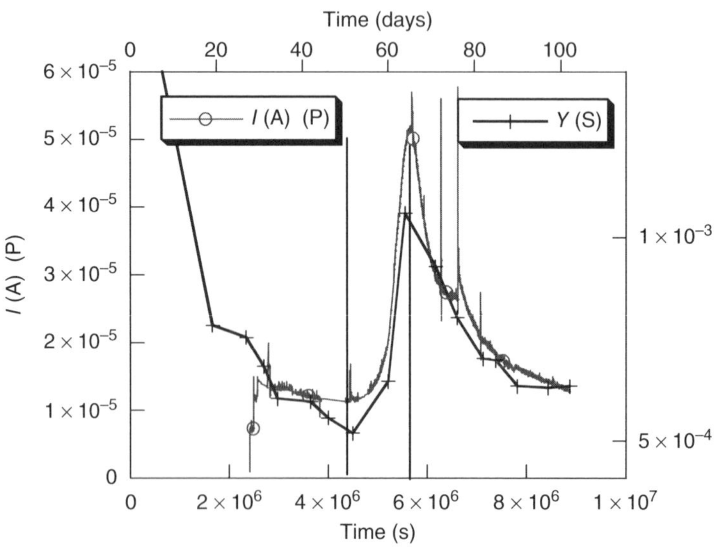

Fig. 7.13 The anodic current and the real conductance variations as a function of time. P, polarised.

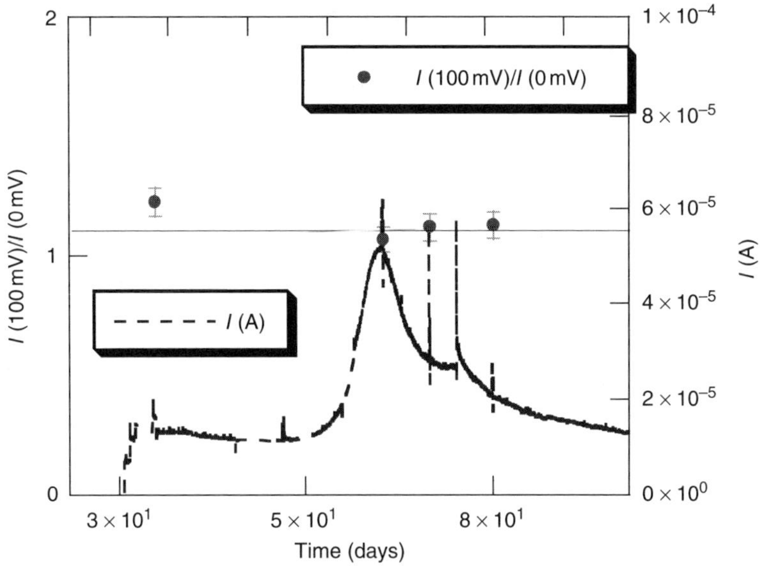

Fig. 7.14 Variations of the anodic current (dashed line) and the rate ratio (round points) variations as a function of time.

at different oxidation times and of determining the current variation due to this modification. If the current ratio, equal to the corrosion rate ratio, is constant as a function of time, the rate-limiting step hypothesis is justified in this case. The variations of this rate ratio are illustrated in Fig. 7.14.

The rate ratio is quite constant during the post-transition. Consequently, the rate-limiting step assumption after the rate peak is justified for this reaction system. Moreover, the current ratio value is around 1.1 for a potential increase of 100 mV, which confirms the slight potential dependence of the corrosion rate. This technique is actually a good way to estimate the *I–E* curve slope on the passivity plateau.

Furthermore, as the rate decreases in the post-transition, it means that it is monitored by a diffusion process, such as the vacancy transport. By assuming that the external interface vacancy concentration is negligible with regard to the internal one and that the electric field in the oxide layer is quite constant and relatively low, the rate ratio for a diffusion process is yielded by the following expression [17, 18]:

$$\frac{I(V_2)}{I(V_1)} = \exp\left((1-\alpha)\frac{F}{RT}(V_2 - V_1)\right) \qquad [7.6]$$

where α is the polarisability of the external interface and V_2 and V_1 are the potential values, 0 and 100 mV versus the reference electrode respectively.

Knowing the value of the current ratio (around 1.1), the polarisability of the internal interface can be evaluated. This parameter $(1 - \alpha)$ is equal to 0.05, which indicates that this interface is not very sensitive to a potential variation.

Impedance spectra under polarisation

As mentioned earlier, the global current density is composed of an anodic part and a cathodic one:

$$i = i_a + i_c \quad [7.7]$$

Thus taking into account these two contributions, the impedance is expressed as follows:

$$\frac{1}{Z} = \frac{\Delta i}{\Delta \eta} = \frac{\Delta i_a}{\Delta \eta} + \frac{\Delta i_c}{\Delta \eta} = \frac{1}{Z_a} + \frac{1}{Z_c} \quad [7.8]$$

with Z_a and Z_c the impedances respectively due to the anodic and cathodic currents.

In order to show the effect of the polarisation, two spectra are exhibited in Fig. 7.15, one obtained with the UP sample and the other acquired at the same oxidation time corresponding to the P sample. As the potential of the P sample is sufficiently anodic, the cathodic current is negligible. As a

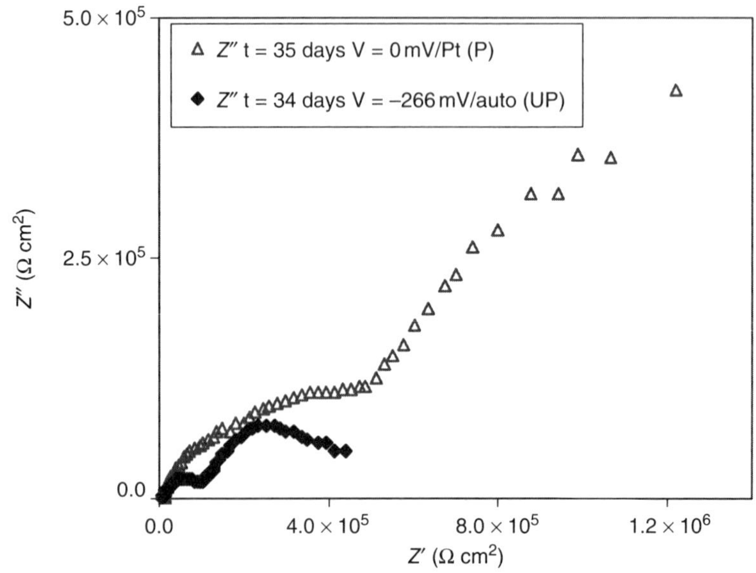

Fig. 7.15 Comparison of the spectra obtained with the sample under polarization (P) and the unpolarised (UP) one.

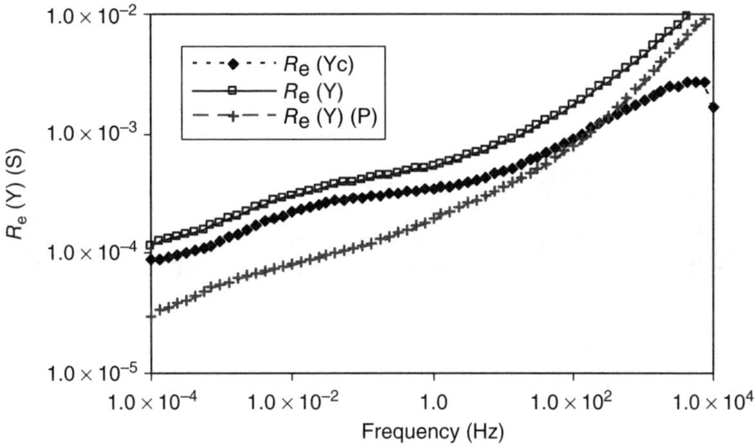

Fig. 7.16 Real part variations of the anodic admittance ($R_e(Y)(P)$), cathodic admittance ($R_e(Y_C)$) and global admittance ($R_e(Y)$) versus Frequency (Hz).

consequence, the admittance ($1/Z$) obtained with this sample only results from the anodic current, whereas that of the UP sample reveals the contributions of the anodic current and the cathodic one. These curves thus indicate that the cathodic process involves a huge decrease in the global impedance. And as its signature is due to the electronic transfer from the metal to the solution through the oxide layer, three steps – namely the charge (electron) transfer from the metal to the oxide, the electronic conduction in the oxide layer and/or the external interface reaction (water reduction) – may be responsible for this decrease. On the other hand, the anodic impedance is indicative of the vacancy diffusion and/or the interface reaction of the production or the consumption of this point defect. At present, we are not able to assign these steps to each signature. However, by supposing that the anodic impedance is equally the same with or without polarisation, it is possible to deduce from the global impedance the cathodic contribution, as shown in Fig. 7.16, which describes the real part of the anodic, cathodic, global admittance with respect to the frequency. According to Fig. 7.16, the cathodic contribution is predominant at low frequencies, whereas in the high frequency domain it is much lower than the anodic contribution.

In conclusion, the global impedance is related to the anodic process at high frequencies and to the cathodic one at low frequencies, as suggested by Sala *et al.* [6].

7.3.2 Sample characterisation

The morphology of the UP sample oxide layer after 103 days of oxidation is illustrated in Fig. 7.17. This shows a relatively uniform oxide layer and a

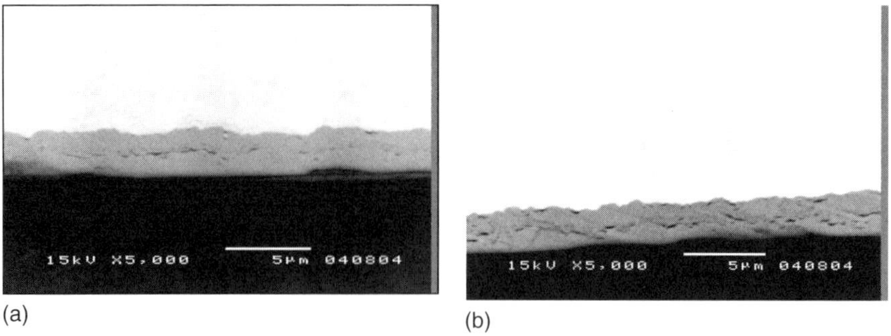

Fig. 7.17 Oxide layer cross-sections formed on the unpolarised sample taken at two different locations: (a) and (b).

quite regular interface. Whichever micrograph is considered, the thicknesses are slightly below 3 µm. A dark line appears at around 1.5 µm from the external interface, which is the result of the oxide layer degradation observed during the kinetic transition. Moreover, the distance between this long crack and the internal interface is variable and depends significantly on the analysed area, which confirms the fact that the transition time is relatively long (15 days).

For the P sample, the conclusions are similar to those obtained with the UP sample. The only difference between these micrographs and the latter ones concerns the oxide layer thickness, which is greater than 3 µm for the P sample (Fig. 7.18). It seems also that there are more cracks at the internal interface on the polarised sample than on the UP sample. The former may be closer to the second kinetic transition than the latter.

In order to evaluate and compare the average thicknesses, more than 20 different areas were analysed with each sample. The results obtained for each sample are presented on the histograms in Fig. 7.19. For the P sample, more than 60% of the analysed area presents an oxide layer thickness between 3 and 3.5 µm, whereas the thickness for the UP sample seems to be less than 3 µm. The average thicknesses determined from these micrographs are actually 2.95 and 3.1 µm respectively for the UP and P samples.

In conclusion, the corrosion rate of Zircaloy-4 in the primary media seems to depend only slightly on the potential. However, it should be noticed that a doubt remains concerning the influence of the polarisation on the corrosion rate. Actually, if the current ratio deduced from Fig. 7.11 (1.3) was constant during the corrosion process, the difference between the P sample thickness and that of the UP sample should be around 0.5 µm, contrary to the mismatch observed by SEM, which is only of 0.15 µm. This discrepancy can be due either to the uncertainty about the thickness determined by SEM or linked to the corrosion rate calculated from the fitting

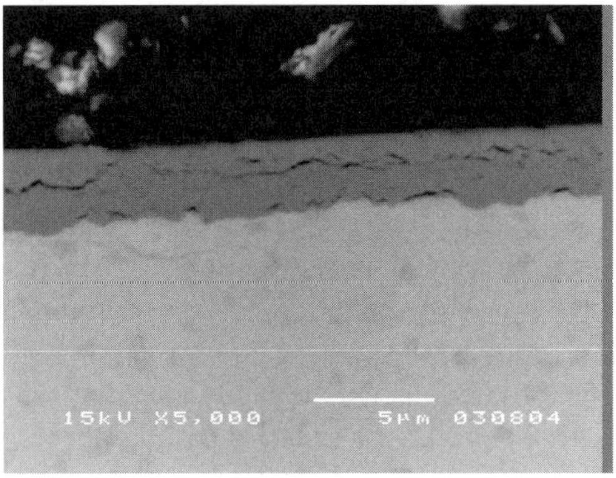

(a)

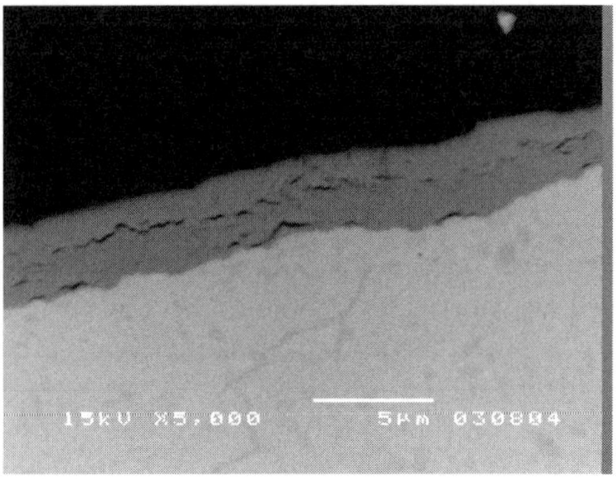

(b)

Fig. 7.18 Oxide layer cross-sections formed on the polarised sample taken at two different locations: (a) and (b).

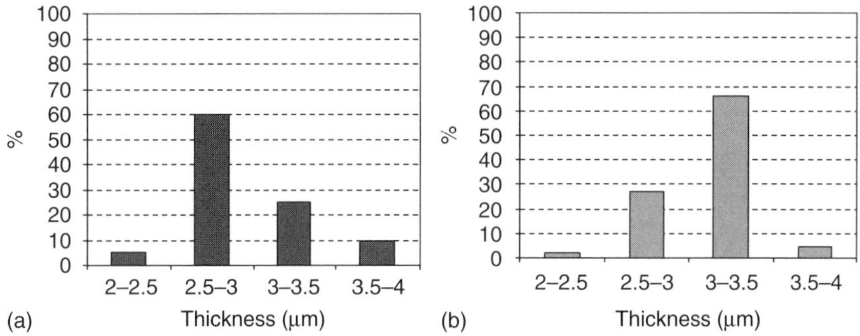

Fig. 7.19 Oxide layer thickness histograms formed on the unpolarised (a) and the polarised (b) samples.

model curve (cf. Fig. 7.11) or to an undesirable current, such as that resulting from the hydrogen oxidation. This phenomenon, however, is not very probable because the oxide is an n-type semiconductor.

7.4 Conclusions

The first aim of this work was to design and construct a practical device allowing us to follow *in situ* the corrosion process by electrochemical measurements. The most important advantage of this original experimental system developed in our laboratory is its simplicity and that the working electrode system used is very close to the fuel rod in a light water reactor.

Moreover, this study has confirmed that the impedance spectroscopy is a good tool to follow the corrosion rate of zirconium alloys, as shown in previous works [6]. The technique allows us to distinguish the different kinetic regimes, such as the pre-transition stage, the kinetic transition and the post-transition stage. However, the determination of the oxide layer thickness from the impedance spectra is not very accurate. Nevertheless, this technique yields a rough estimation and allows us to follow oxide growth on-line.

Because of this uncertainty, one of the samples was submitted to an anodic polarisation, in order to access directly to the anodic current. In this sense, the corrosion process of this polarised sample was followed by chronoamperometry. Contrary to the impedance spectroscopy, this technique gives directly the oxidation rate as a function of time.

This study shows also that the corrosion rate depends only slightly on the potential, which proves that the free corrosion potential is located on the passivity plateau. Moreover, it seems that the currents of the UP and the P samples are similar during the pre-transition stage, and even beyond, but this must be confirmed by future work.

The impedance spectra obtained with the P sample are clearly above those acquired at the open circuit potential. And these are representative of the anodic process whereas both the anodic and cathodic currents contribute towards the impedance obtained with the UP sample. By assuming that the anodic impedance is independent of the potential, the cathodic impedance can be deduced from the global one. The dissociation of these contributions shows that the anodic and cathodic processes are respectively predominant in high and low frequencies. The most difficult thing is to assign each signature to a physico-chemical process.

Finally, this study proves once again that impedance spectroscopy is a good tool to follow the corrosion process of Zircaloy-4 but underlines also that it is important to couple this with another technique such as chronoamperometry.

7.5 References

1. D. Pecheur, J. Godlewski, P. Billot and J. Thomazet, *Proc. 11th Int. Symp. Zirconium in the Nuclear Industry, ASTM STP 1295*, 1996, 94–113.
2. P. Bossis, G. Lelièvre, P. Barberis, X. Iltis and F. Lefebvre, *Proc. 12th Int. Symp. Zirconium in the Nuclear Industry, ASTM STP 1354*, 2000, 918–945.
3. T. Isobe, Y. Hatano and M. Sugisaki, *J. Nucl. Mater.*, 1997, **248**, 315.
4. P. Rüdling, *J. Nucl. Mater.*, 1999, **265**, 44.
5. M. Tupin, M. Pijolat, F. Valdivieso, M. Soustelle, A. Frichet and P. Barberis, *J. Nucl. Mater.*, 2003, **317**, 130–144.
6. B. Sala, I. Bobin-Vastra and A. Frichet, *Proc. 8th Int. Symp. Environmental Degradation of Materials in Nuclear Power Systems – Water Reactors*, Amelia Island, Florida, 1997, Vol. 2, 1005.
7. J. Schefold, D. Lincot, A. Ambard and O. Kerrec, *J. Electrochem. Soc.*, 2003, **150**(10), B451–B461.
8. M. Parise, PhD Thesis, Ecole National supérieure des Mines de Paris, 1996.
9. G. Beranger and P. Lacombe, *J. Nucl. Mater.*, 1965, **16**, 190.
10. G. P. Sabol and S. B. Dalgaard, *J. Electrochem. Soc.*, 1975, **122**(2), 316.
11. J. Godlewski, J. P. Gros, M. Lambertin, J. F. Wadier and H. Weidinger, *Proc. 9th Int. Symp. Zirconium in the Nuclear Industry, ASTM STP 1132*, 1991, 416–436.
12. C. Gabrielli, 'Identification of electrochemical processes by frequency response analysis', technical report number 004/83, Solartron Schlumberger, 1984.
13. C. Soustelle and M. Pijolat, *Solid State Ionics*, 1997, **95**, 33.
14. K. Surla, F. Valdivieso, M. Pijolat, M. Soustelle, M. Prin-Lamaze, *Solid State Ionics*, 2001, **143**, 355.
15. A. K. Jonscher, *Dielectric Relaxation in Solids*, Chelsea Dielectrics Press, London, 1983.
16. C. Bataillon and S. Brunet, *Electrochim. Acta*, 1994, **39**, 455.
17. A. T. Jr. Fromhold, *Theory of Metal Oxidation*, Space Charge/North-Holland, New York, 1980, Vol. 2.
18. C.-Y. Chao, L. F. Lin and D. D. Macdonald, *J. Electrochem. Soc.*, 1981, **128**, 1187–1194.

8
Electrochemical corrosion study of Magnox Al80 and natural uranium

R. BURROWS and S. HARRIS, Nexia Solutions Ltd, UK

8.1 Introduction

Spent fuel elements from the UK Magnox nuclear power station fleet are stored in cooling ponds to allow decay of short lived radioisotopes. The high magnesium alloy Magnox Al80 is used as clad material due to its low neutron absorption cross-section. Clad penetration during aqueous storage would expose the uranium fuel to the pond water and potentially result in the release of fission products through uranium corrosion.

Magnesium and its alloys quickly passivate in alkaline conditions due to the formation of a magnesium oxide/hydroxide layer [1]. This film has a high stability due to its low solubility and rate of dissolution but can be disrupted by the action of aggressive anions, most notably chloride, which promote localised corrosion [2]. It is well known that the general corrosion rates of Magnox in alkaline conditions are sufficiently low that no penetration of fuel element clad would be expected to occur over time scales of many years. However, localised attack can produce corrosion rates of the order of many millimetres per year in isolated areas. A number of dynamic processes are involved during Magnox corrosion under alkaline conditions, the most important being passive layer formation and the carbonation of the electrolyte.

Uranium corrodes in an active–passive manner, similar to that exhibited by stainless steels [3]. The rate at which corrosion proceeds is influenced by the degree of irradiation swelling which occurs as a result of fission product precipitation within the uranium fuel matrix [4]. Temperature also influences reaction rate; however, there have been few studies of the effects of solution composition under cooling pond conditions.

Fuel element aqueous storage environments have closely controlled chemistry; dosing of sodium hydroxide maintains a high alkalinity and the effect of absorption of atmospheric carbon dioxide is minimised by ion exchange treatment, which similarly maintains a low anion impurity level. The combined chloride and sulphate concentration is typically significantly below $1\,g\,m^{-3}$ [1].

Substantial research work has been undertaken on the corrosion behaviour of Magnox Al80 and reactor grade uranium by the former CEGB [1, 4, 5], and more recently by BNFL. Performance under nominal conditions is well understood as illustrated by several decades of plant operation. Corrosion monitoring techniques have been developed significantly, however, and the application of these contemporary methods is prudent in order to allow as full an understanding to be developed as possible. Electrochemical methods have been utilised historically in the study of Magnox [2], uranium [3] and more recently, with automotive magnesium alloys [6, 7] and these are suitable for studying what is a peculiar corrosion behaviour with enhanced sensitivity and depth of mechanistic detail. Here, contemporary methods have been employed to gain greater insight into the corrosion behaviour of Magnox Al80 and natural uranium under aqueous storage conditions, with particular regard to pitting initiation.

8.2 Method

Electrochemical tests employed a standard electrochemical cell utilising a three-electrode geometry. Working electrodes comprised commercial purity material and were of $2\,cm^2$ surface area and, in the case of Magnox electrodes, were etched with 2% citric acid immediately prior to use in order to minimise surface film formation. Potentiodynamic and AC impedance tests were performed with the use of Solartron 1280B Electrochemical Measurement Units. Data collection and post-processing involved the use of the Scribner Associates suite of electrochemistry programmes. Tests employed an Ag/AgCl reference electrode and platinum counter-electrode.

Electrolytes were prepared from high purity deionised water and AnalaR grade reagents. The nominal solution chemistry was $200\,g\,m^{-3}$ sodium hydroxide which was equivalent to pH 11.7 at 25°C. The pH was measured at intervals with the use of an Orion pH probe and meter.

Potentiodynamic scans were generally made in the range –0.5 to +3.0 V vs open circuit potential (OCP) with a scan rate of $0.2\,mV\,s^{-1}$. AC impedance involved the application of a 10 mV excitation at OCP over the frequency range 2×10^5 to 0.1 Hz, with analysis of the current response by the Solartron 1280B internal frequency response analyser. The corrosion rate was calculated by analysis according to a simple equivalent circuit [8].

8.3 Results: Magnox Al80

Potentiodynamic tests demonstrated clearly the formation of a stable passive film on Magnox Al80 surfaces at pH 11.7 and above, with strongly limited anodic current densities (Fig. 8.1 and 8.2). Similarly, AC impedance

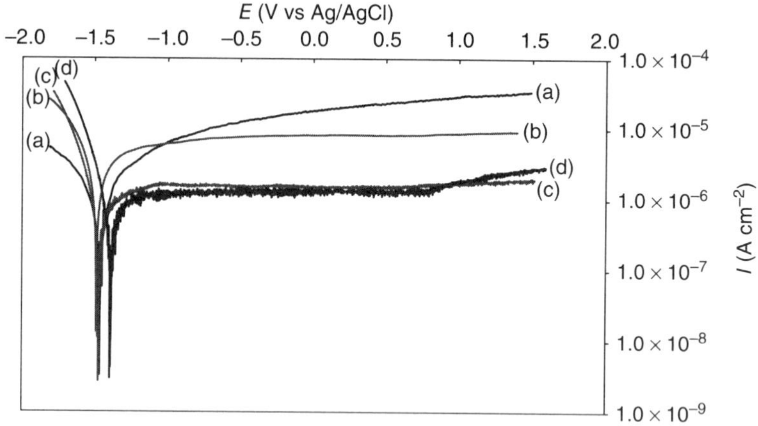

Fig. 8.1 Anodic polarisation of Magnox in aqueous solutions of pH 10.0 (a), 11.0 (b), 11.7 (c) and 12.0 (d).

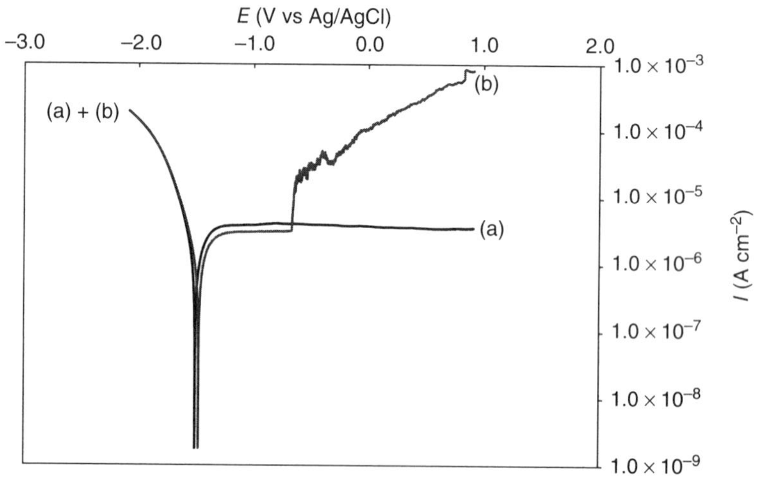

Fig. 8.2 Anodic polarisation of Magnox in pH 11.7 without (a) and with (b) 10 mg kg^{-1} chloride.

measurements showed that the corrosion rate fell very quickly to a low rate in the hours following immersion (Fig. 8.3–8.7). The effect of low levels of chloride was also apparent and initiation of localised corrosion was observed as a sudden increase in current density in a Magnox sample exposed to 10 mg kg^{-1} chloride (Fig. 8.2). Initiation of localised corrosion could be verified visually by the evolution of hydrogen on the sample surface. Prior to this breakdown in passivity, the electrochemical behaviour of the Magnox surface was indistinguishable from the sample in the absence of chloride.

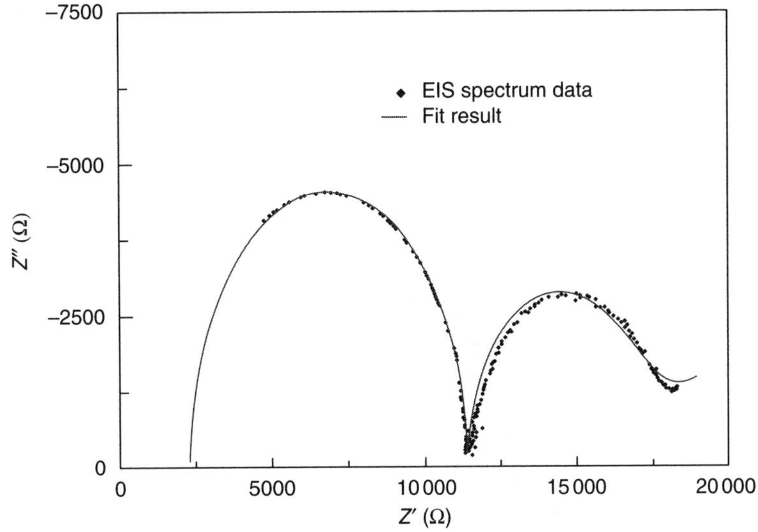

Fig. 8.3 Sample Magnox Al80 impedance spectrum with fitting from model in Fig. 8.4. EIS, electrochemical impedance spectroscopy.

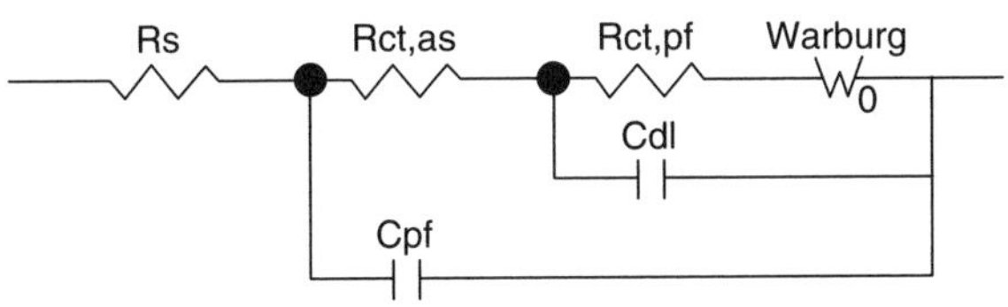

Fig. 8.4 Equivalent circuit model used for fitting Magnox a.c. impedance spectra.

Although the use of d.c. techniques yielded much useful information, the dynamic nature of the system under study has highlighted the shortcomings of these methods, as they demand a steady state system; a.c. impedance spectra may be collected over far shorter periods and therefore allow the initial passivation processes occurring on the Magnox Al80 surface to be studied.

The following common features were exhibited in Magnox impedance spectra (Fig. 8.3–8.7):

- A large semicircular time constant representing the charge transfer across the passive layer at medium frequency.
- A second smaller time constant representing a second (possibly localised) corrosion process at high frequency.
- A Warburg diffusion impedance (constant gradient tail) at low frequency.

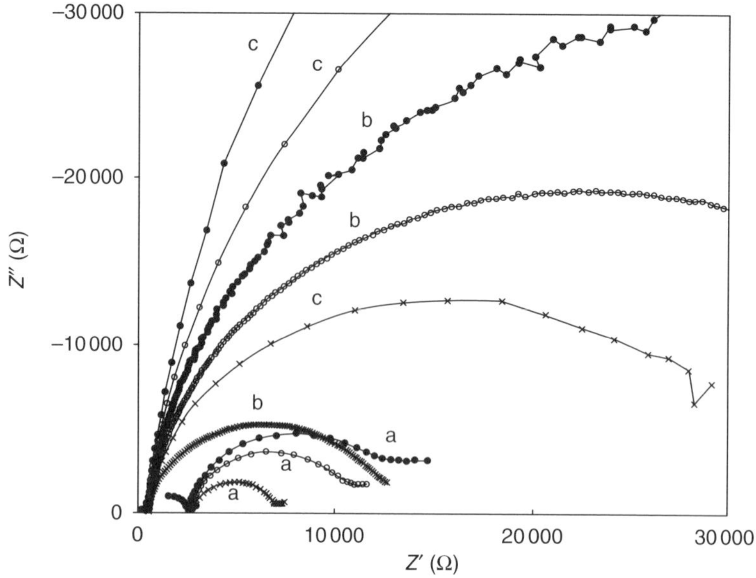

Fig. 8.5 Impedance spectra of Magnox in pH 11.0 (a), 11.7 (b) and 12.0 (c) after 1 h (×), 5 h (○) and 10 h (•).

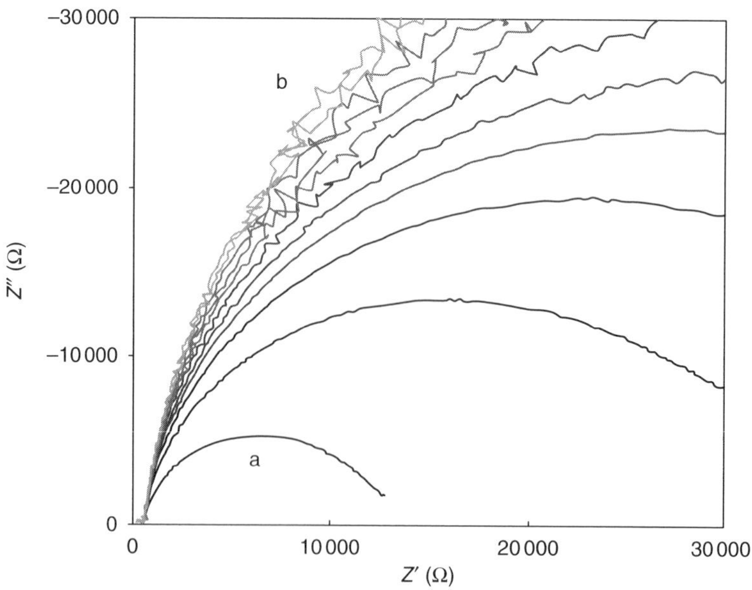

Fig. 8.6 Impedance spectra of Magnox in pH 11.7 from 1 h (a) to 20 h (b).

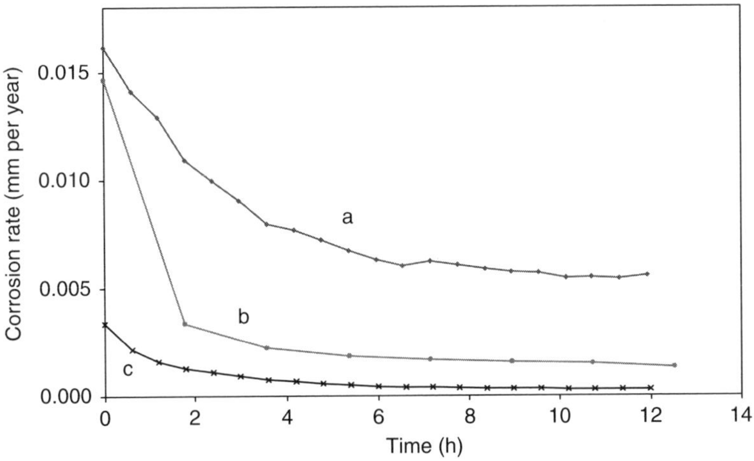

Fig. 8.7 Corrosion rates of Magnox in pH 11.0 (a), 11.7 (b) and 12.0 (c).

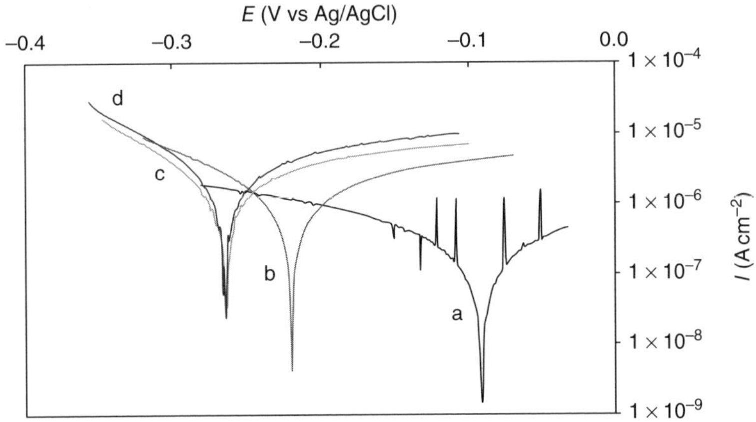

Fig. 8.8 Anodic polarisation of uranium in solutions of pH 10.0 (a), 11.0 (b), 11.7 (c) and 12.0 (d).

8.4 Results: natural uranium

The results of anodic polarisation tests show that increasing the electrolyte pH results in a cathodic shift of the E_{corr} value and an increase in the corrosion current density (Fig. 8.8). a.c. impedance spectra are simpler than Magnox, possessing a single time constant and a Warburg diffusion impedance which is apparent at the lower frequency limit (Fig. 8.9 and 8.10). The calculated a.c. impedance corrosion rates are in good agreement with the d.c. results, particularly for the lower pH experiments (Fig. 8.11).

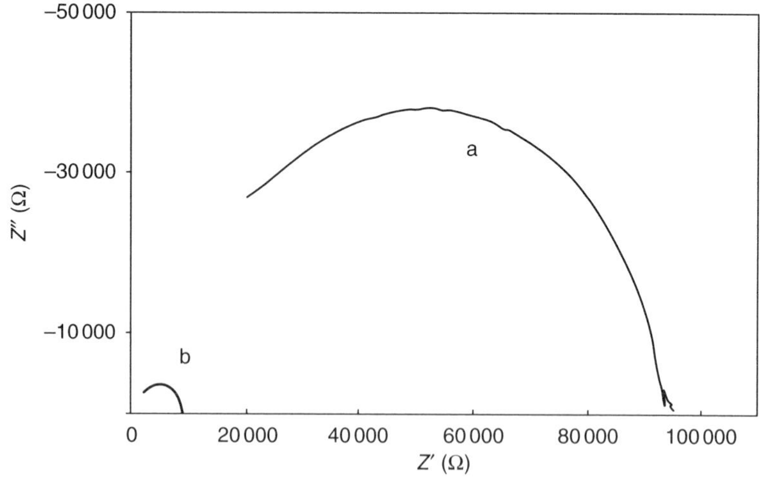

Fig. 8.9 Impedance spectra of uranium in pH 10.0 (a) and 11.0 (b).

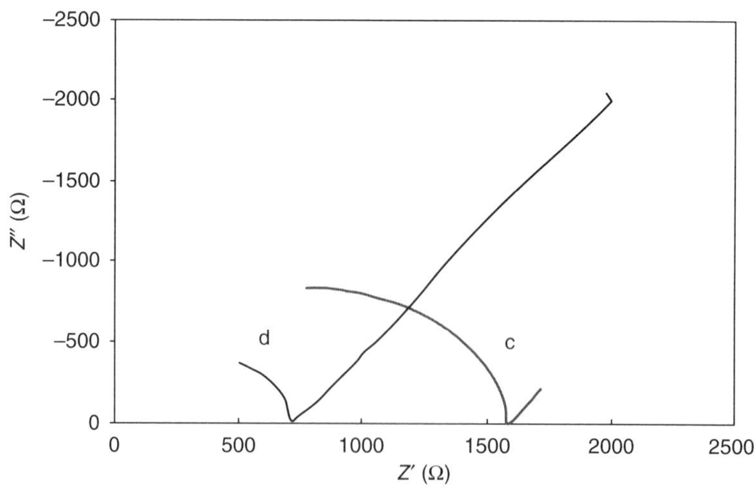

Fig. 8.10 Impedance spectra of uranium in pH 11.7 (c) and 12.0 (d).

8.5 Conclusion

The employment of a.c. impedance spectroscopy enabled the study of the initial stages of Magnox passive film formation, showing that in alkaline conditions, a passive layer formed on the electrode surface resulting in diffusion-limited corrosion behaviour. The presence of chloride promoted the breakdown of this passive film at high anodic potentials. The performance of Magnox clad as a fuel clad material is dependent upon its ability to resist corrosive attack which can lead to localised pitting.

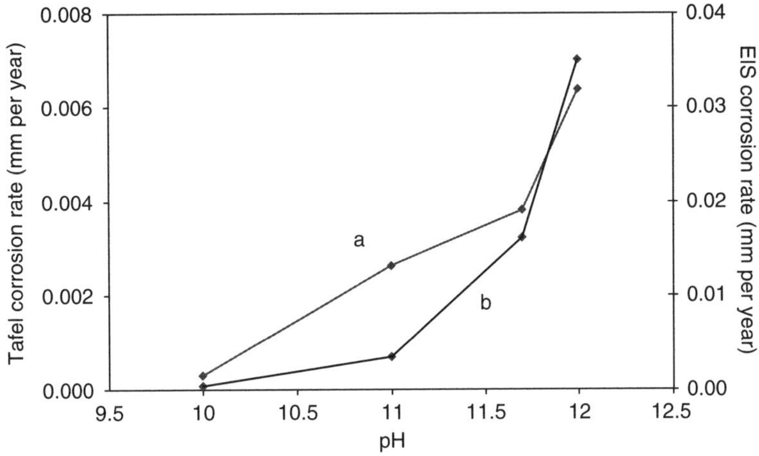

Fig. 8.11 Comparison of natural uranium corrosion rates from Tafel (a) and impedance (b) data. EIS, electrochemical impedance spectroscopy.

The impedance spectrum of Magnox exhibited two time constants and a Warburg feature indicative of a diffusion-limited mechanism. Uranium corrosion exhibited an active–passive behaviour, consequently exhibiting a simple impedance spectrum.

Complementary use of AC impedance spectroscopy alongside d.c. polarisation has allowed the study of general and localised corrosion, with greatly enhanced sensitivity and reduced time scales. Significant practical benefits have also been afforded; for example, the radiological wastes arising are reduced by shorter measurement times.

8.6 References

1. Bradford P. M., Case B., Dearnaley G., Turner J. F., Woolsey I. S., *Corr. Sci.*, **16** (1976) 747.
2. Thomas M., Weber J., *Werkstoffe Korr.*, **19**(10) (1968) 895.
3. Ward J. W., Waber J. T., *J. Electrochem. Soc.*, **109**(2) (1962) 76.
4. Tyfield S. P., *Nucl. Energy*, **27**(2) (1988) 91.
5. Friskney C. A., *J. Nucl. Mater.*, **99** (1981) 165.
6. Song G., Atrens A., St John D., Wu X., Nairn J., *Corr. Sci.*, **39**(10–11) (1997) 1981.
7. Baril G., Blanc C., Pebere N., *J. Electrochem. Soc.*, **148**(12) (2001) B489.
8. Walter G. W., *Corr. Sci.*, **26**(9) (1986) 681.

Part III

Electrochemistry and tribocorrosion issues

9

Analysis of mechanisms induced by sliding and corrosion: dedicated apparatus for pressurized water reactor environments

J-PH. VERNOT and D. KACZOROWSKI,
AREVA NP, France

9.1 Introduction

In pressurized water reactors (PWRs), some components are submitted to relative motions due to operational processes (localization and positioning adjustment) or to undesired effects (flow induced vibration). Thus, components and associated supports are typically excited by a large range of kinematics so that complex combinations of wear can occur. Furthermore, for each operational process [e.g. control rod drive mechanism (CRDM), rod cluster control assembly (RCCA), fuel rod in the assembly, etc.] these components have a very specific geometry leading to a very low contact pressure or a very high contact or pressure (typical impact and sliding for latch arm drive mechanism). Materials used in this application are usually austenitic stainless steel or zirconium alloys.

To understand the wear behaviour of the two partners in pressurized high temperature water, a tribometer able to work under this specific condition is used. The excitation movement is adjusted with regard to measurements and analyses obtained on real components in the plant or derived from simulators at different scales. More specific analyses between wear facies in the plant and damage on the samples in the tribometer are also completed to ensure that the degradation process is governed by the same main wear phenomena. This comparative examination revealed that degradation cannot be explained only by mechanical aspects but must be associated with other aspects like corrosion activity.

Let us recall that typical PWR environments coupling temperature (320°C), pressure (154 bar) and chemistry solution (deaerated, low conductivity water) involve corrosion processes which can play an important role in the development and synergy of wear processes. It has been clearly identified that wear due to mechanical aspects and wear due to corrosion aspects are lower than the wear when the two components are coupled [1, 2]. It has been shown for CRDMs, that interactions between drive rod of AISI 410 martensitic stainless steel and gripper latch arm protected by a cobalt-based hard facing alloy (Stellite grade 6), have a large effect of time latency on the degradation [3]. The comparison consists on the one hand of results acquired from the mock-up scale 1, and on the other hand of the experience feedback on the components. The main difference was that the interaction has been reproduced at a frequency greater than that applied on plant. Due to this time latency effect, the wear damage is lower for high frequency interaction and can introduce singular artefacts to estimate the lifetime of components [3].

These kinds of experiences highlight thc importance of corrosion processes in PWR conditions such as those clearly shown by a parametric study carried out at room temperature with activated water (e.g., NaCl, H_2SO_4, etc.) [4]. Nevertheless tribocorrosion has been studied on analytical systems such as the pin-on-disk tribometer and mainly with typical environments which probably overestimate the influence of corrosion. That is why it seems difficult to extrapolate the results to PWR environments.

To investigate the repassivation phenomenon in such particular environments, Framatome ANP has developed a specific apparatus running at PWR conditions in order to understand the complex synergy when mechanical and corrosion effects are occurring at the same time. The aim of this investigation is to distinguish the relative influence of corrosion in wear processes considering that a number of parameters can be involved, i.e., materials, surface preparation and roughness, normal load, kinematics, time between contacts, temperature, chemistry of the environment, etc.

9.2 Experimental device

9.2.1 Description of the original apparatus

The test apparatus was originally designed to test under PWR conditions two contacting surfaces such as control rod against support, fuel rod section against grid. The original test bench, dedicated to mechanical wear processes, is housed in an autoclave at a temperature up to 320°C ± 5°C and a pressure of 154 bar. The water environment during the test is preliminarily controlled and its chemical composition is similar to the primary coolant PWR ($[O_2] < 0.1$ ppm, $[B] = 1300$ ppm, $[Li] = 2$ ppm, $[Cl^-] = 0.1$ ppm), its pH is about 7 and the bulk is deaerated [5]. Nevertheless, composition can be adjusted to particular operating conditions.

A schematic representation of the tribometer is presented in Fig. 9.1. The samples are a tube element and a ring. The tube is moved in the autoclave by a specific excitation induced by four electromagnets each oriented at 90° to each other. Figure 9.2 represents the trajectories of the centre of the tube

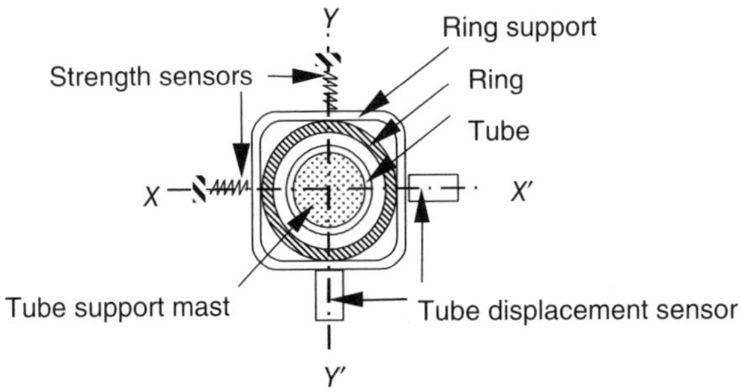

Fig. 9.1 Schematic cross-section of the tribological apparatus.

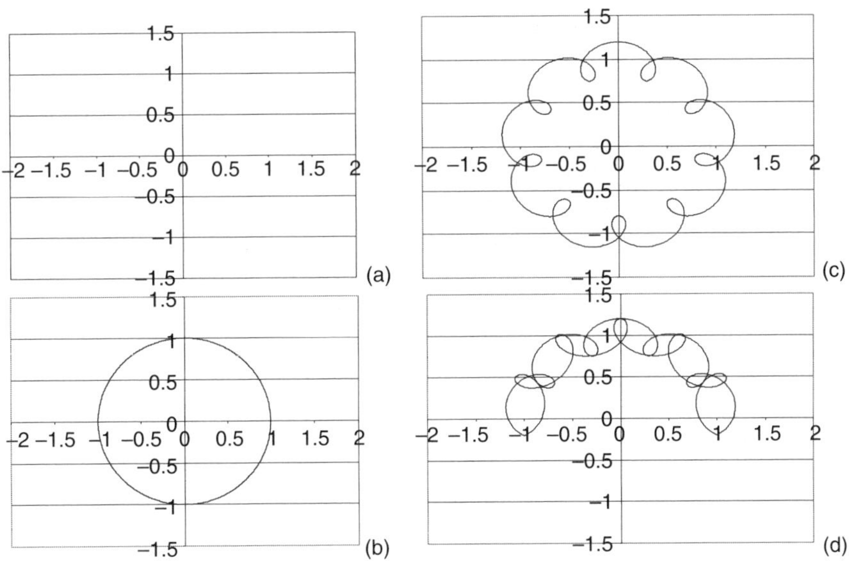

Fig. 9.2 Trajectories of the centre of the tube for various kinematics. (a) Normal impact: movement is rectilinear and the impact zone is always the same. (b) Rolling with orbital sliding: the tube's centre rotates on a 500 μm radius circle. (c) Impact plus orbital sliding. (d) Impact plus semi-orbital sliding, the tube moves forward and backward.

for several combinations of electromagnet piloting. Typical movements can be proposed to induce solicitations as simple impacts, orbital sliding, semi-orbital sliding, orbital sliding plus impact, semi-orbital sliding plus impact or fretting.

The test specimens are cylindrical stationary sleeves and cylindrical tubes (diameter of tube = 9.7 mm, diameter of sleeve = 10.7 mm).

- *Sheer impact* is used to obtain a normal contact of the tube with the ring.
- *Sliding with an orbital movement* is detailed in Fig. 9.3. During the movement, the contact point C makes revolutions, at a frequency f_1 between 1 to 10 lap s^{-1}, on the tube and on the ring. Therefore, it is a rolling and sliding movement.
- *Impact plus sliding with an orbital movement* is performed by superposing a higher frequency, f_2 (5–120 Hz), to the orbital sliding described above. The contact between the tube and the sleeve is thus periodically disturbed. The result is a succession of oblique impacts with sliding. Signals composed of two frequencies are applied to electromagnets in order to obtain these kinematics.
- *Impact plus sliding with a semi-orbital movement* is also encountered in reality. It is obtained in the test bench when impacts are spread over only half of the periphery of the specimens.

In this specific tribometer, forces and displacements are recorded during the test. The trajectory of the tube support mast is measured by eddy current sensors. The deflection measurement of the ring support gives the contact force between the specimens.

The analysis of each channel measurement, with dedicated software running at high frequency, allowed us to correlate normal and tangential displacements, normal and tangential forces. Statistical analyses are also performed in order to characterize the evolution of main parameters (wear evolution, friction coefficient, duration of the contact, sliding distance, etc.). At the end of the test, results are compared to local investigation performed on the samples by several methods (microscopic analysis, SEM observation, etc.).

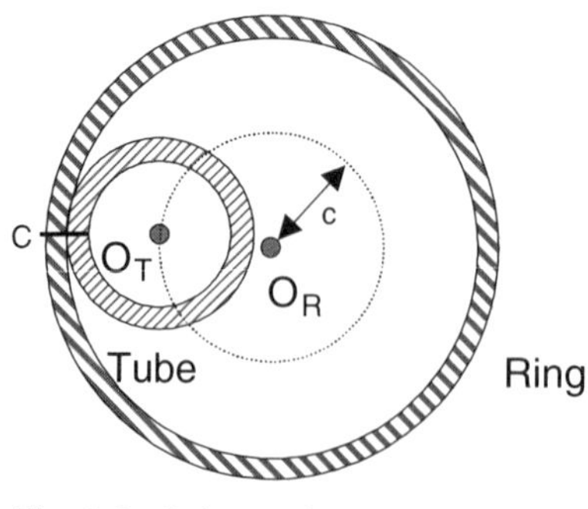

Fig. 9.3 Schematic representation of the orbital rolling and sliding. Contact is maintained between ring and tube at the point C. An orbital motion is applied to the centre of the tube (O_T). Due to the radius difference between tube and sleeve, a relative sliding speed occurs between the two specimens.

It has been shown in previous studies [6] that contact kinematics is an important parameter in order to represent the same wear scars as obtained in a power plant.

9.2.2 Description of the adapted apparatus

Based on the well-known mechanical test bench described above for specific mechanical measurement at PWR conditions, electrochemical measurements were implemented in order to characterize corrosion effects; in particular to define the reciprocal influence of the mechanical and electrochemical parameters during the test: force, displacement, current, potential in connection with parameters like water chemistry, time latency, contact forces, etc.

Eddy current sensors are used for the first measurements. They give mechanical information while electrochemical information about the bulk influence and oxidation reactions are given by current and potential which are measured in relation to an Ag/AgCl main reference electrode or to a secondary reference.

In fact, because of the dimensions of the actual apparatus the Ag/AgCl reference electrode is too voluminous to be put near the specimens. Consequently, as this electrode is essential, it is put on the side of the apparatus but quite far from the specimens. This leads to signal disruption problems due to ohmic fall and answer quickness. This is avoided by the use of a secondary reference placed as near as possible to the samples.

The Aurore apparatus principle is described in Fig. 9.4. The principle of electrochemical measures is the classical one with three electrodes. The reference electrode is Ag/AgCl, the working electrodes are the specimens subjected to rubbing in the specific bulk described above. The counter-electrode also is a noble metal very similar to thc secondary reference electrode.

The samples were defined as small as possible but providing a good representation of the contact phenomenon in a higher scale in order to decrease the galvanic coupling risk. Dimensions were improved, e.g. the only free part is the contact part. It is necessary to electrically insulate the various electrodes, which is done by using zirconium parts. Specific insulation has also been developed for necessary connection and allows leak tightness from the PWR environment to room temperature, all along the linkage to the specimens and electrode.

Another step in the apparatus development had been the upgrade of the software in order to acquire several data on the same computer at the same time. Records come from a Gamry potentiostat for electrochemical measurement and from a data acquisition board for mechanical measurements. The Gamry potentiostat also permitted adjustment of the current or the potential for the test (running in galvanic or in potentiostatic modes).

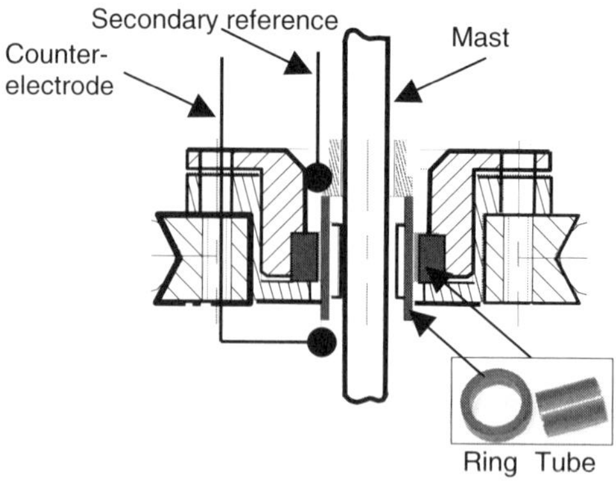

Fig. 9.4 Schematic cross-section of the Aurore apparatus with the three-electrode set-up.

A testimony sample is also used. In contact with the bulk and the solution perturbations due to turbulences created by the movement of the samples, it can be compared with the specimens submitted to rubbing in the bulk.

Wear is calculated from the difference between the weighings carried out before and after the test. In fact the implementation of electrochemistry measurements needed specifics studies, developments and tests that are not described in this chapter. The list includes (and it is not exhaustive): position of the counter-electrode, response on counter-electrode and Ag/AgCl reference, and electromagnet influence.

9.3 Results and discussions

9.3.1 Static characterization

The purpose of the first campaign was to assess the ability of the Aurore electrochemical system to detect the elementary phenomenon of depassivation.

The first set of experiments allowed determination of the polarization curves for the 304L stainless steel ring at 80°C and 300°C. The tests were performed with the Ag/AgCl reference electrode and with two potential scanning speeds: 0.5 and $2\,mV\,s^{-1}$. The influence of the surface state was clearly demonstrated by applying several prepolarizations in the cathodic field before the potential scanning. As shown by Fig. 9.5 an activity peak is observed after a prepolarization of –1 V for 3 min (reference Ag/AgCl) and increases for the same duration at –2 V (for a same voltage scan speed of 2 mV/s). This effect is not so evident at 300°C (Fig. 9.5 and 9.6). A scanning speed effect is noticed due to small differences between the curves obtained at $0.5\,mV\,s^{-1}$ and $2\,mV\,s^{-1}$ (see Fig. 9.5).

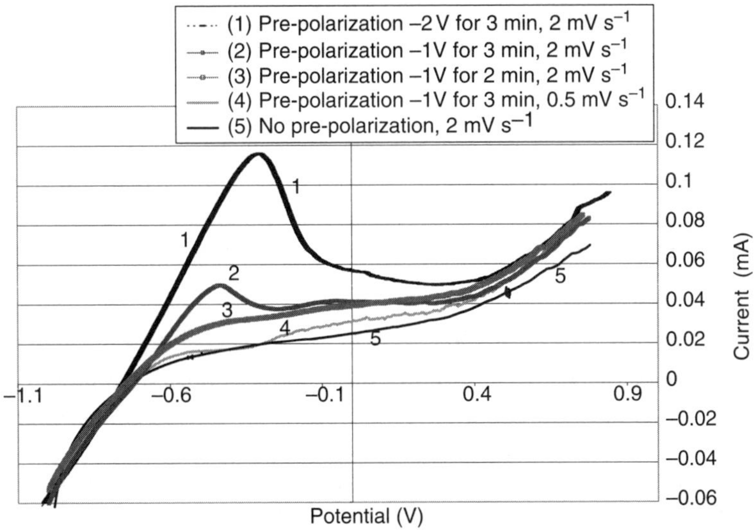

Fig. 9.5 304L stainless steel in pressurized water reactor water chemistry at 80°C, polarization curves for different scanning speed values and various amplitudes and duration of cathodic first polarization; reference electrode is Ag/AgCl; direct (increasing potential) scans.

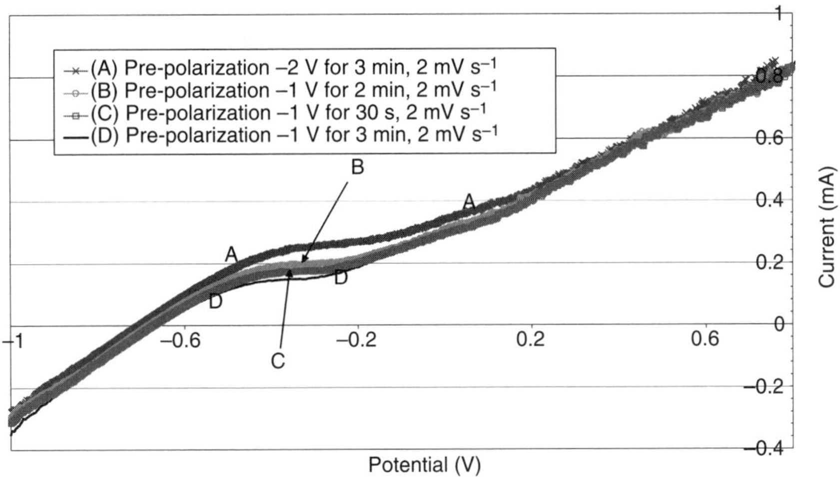

Fig. 9.6 304L stainless steel in pressurized water reactor water chemistry at 300°C, polarization curves for 2 mV s^{-1} scanning speed and various amplitudes and duration of cathodic first pre-polarization; reference electrode is Ag/AgCl; direct (increasing potential) scans.

The result shows a very good reproducibility of the response with a very similar corrosion potential for all the polarization curves. The effect of scanning speed seems to be limited. It appears that due to differences in the initial state (duration and level of the initial potential before scanning), a singular response in the polarization curve such as peak activity is appearing. This can be explained by the high level and the high duration of prepolarization which induce a replenishment of the surface and produce an increasing current before renewal of the substrate.

9.3.2 Scratch test solicitation

The objectives are to determine the electrochemical components of wear in high temperature water. The ability to detect some depassivation phenomena was investigated, with the help of a scratch test. This method creates a single event and allows quantification of the surface area of a scratch generated by an indenter and is suitable to follow the repassivation kinetics of a bare metal surface. This method is commonly used, and some papers deal with its application in pressurized high temperature water [7, 8].

The purpose was to study transitional currents associated with scratches for different polarization values. An indenter described in Fig. 9.7, made of ZrO_2, was developed and adapted to our apparatus as the mobile sample. The trajectory followed by the indenter is described in Fig. 9.7. The static sample is an AISI 304 stainless steel with flat geometry (see Fig. 9.8).

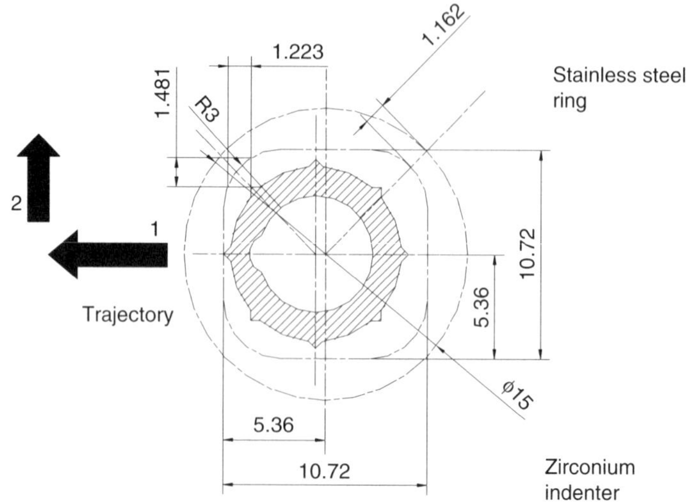

Fig. 9.7 ZrO_2 indenter and trajectory in the ring to create the scratch. (Dimensions are in mm.)

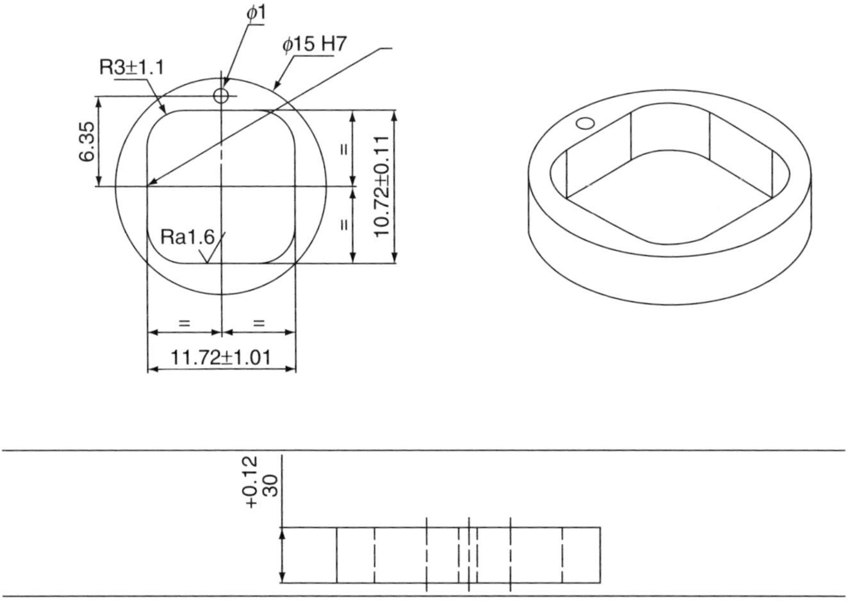

Fig. 9.8 304L stainless steel ring. (Dimensions are in mm, roughness in μm.)

The first test series was performed in 40°C deaerated water and various values of polarization were applied to the ring in order to record transitional current linked to depassivation. The corrosion potential was about –500 mV measured with Ag/AgCl reference electrode. The system appeared to be very sensitive to disruptions. Furthermore the current sent through the magnets generated an important interference signal.

To observe phenomena as Stellwag [7] did, the polarization values have to be very high (polarization of +300 mV higher than corrosion potential). The current transient observed is shown in Fig. 9.9. The applied potential does not vary during the test. Before the scratch occurs, a 40 μA current is applied through the counter-electrode to maintain the polarization. A current peak (amplitude near to 23 μA) is observed early at the beginning of the scratch (the scratch is realized in 10 ms and its length is around 500 μm, according to the displacement signal analysis). The repassivation phenomena is also observed with a duration of 16 s to obtain the current stability. Small electric disturbances are observed on the potential and current signals.

A second series of tests was realized at 300°C under the same conditions as the tests at 40°C. No signal was recorded during the scratch. It seems that the corrosion phenomenon plus the structure interference gives a background noise around 100–150 μA. So, if the transitional current peak is only a few microamperes high it cannot be observed. This battery of tests showed the possibility of realizing polarization curves at 300°C and 154 bars with

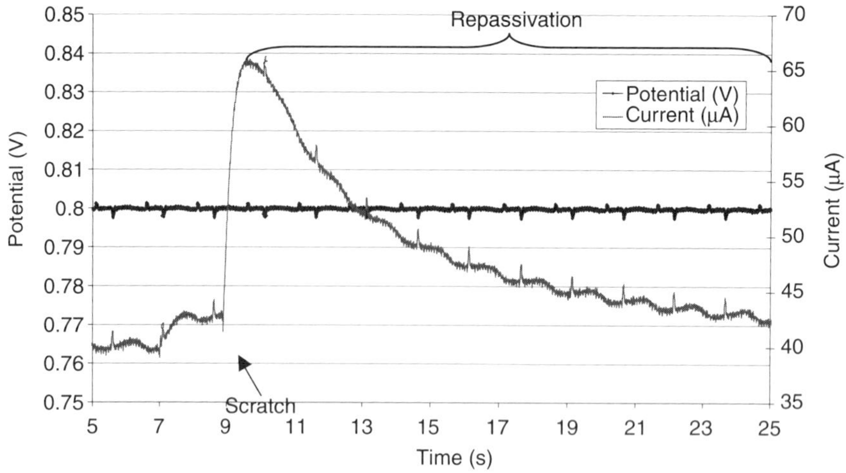

Fig. 9.9 Current evolution during a 40°C scratch test with a +300 mV polarization. The reference electrode is the secondary one.

the existing Aurore system. The initial surface state influences the electrochemical response.

9.3.3 Orbital test configuration

Scratch tests were performed to illustrate the electrochemical signature of an elementary depassivation phenomenon but they were not conclusive at 300°C. The lightly depassivated surface generates a weak answer against the background noise due to fluid movements or magnets. Even though the lower temperature scratch tests seem more useful they mainly show the magnet disruptions. The more highly depassivated surfaces were then investigated. A stainless steel tube was used as a mobile sample and a classical ring (without flat geometry) as a static sample, similar to those described by Bosch *et al.* [8]. The kinematics was rolling with sliding. Both samples are electrically connected at the outside of the tribometer and correspond to a unique working electrode. A secondary reference electrode was used. A polarization +30 mV above the corrosion potential was applied to the samples. A current peak is observed early at the beginning of the friction (see Fig. 9.10) and a stabilized state is obtained after 300 s. At the end of the friction (total duration: 600 s) a decay is observed and the repassivation is observed after 7000 s. The stable state corresponds to the sum of the newly active area created by friction and repassivation.

The total depassivated area concerned the tube and the ring, and was around 2 cm^2, corresponding to the surface affected by sliding. It is clear that this whole surface was not activated at the same time.

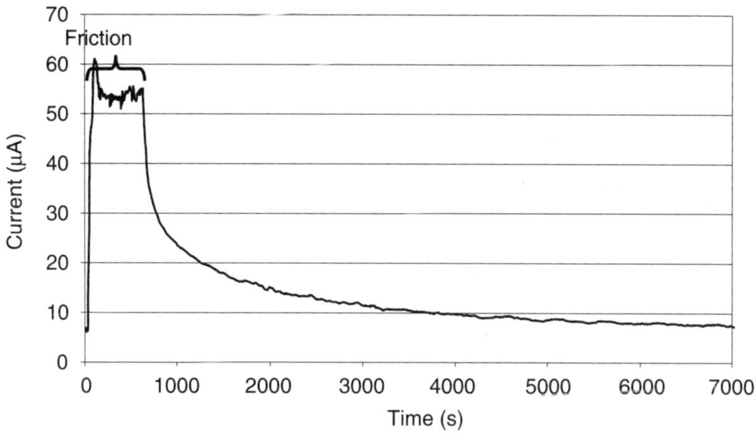

Fig. 9.10 Current transient during orbital sliding friction between two 304L samples. After 600 s, the contact stop and the current decrease. Stability is obtained after 7000 s.

9.4 Conclusions

In PWRs, components can be submitted to a coupling effect involving interactions at localized areas (mechanical aspect) and also the environment atmosphere (corrosion effect). Due to this particular phenomenon, so-called tribocorrosion, it seems difficult to appreciate the relative effect of each component, although studies at room temperature and room pressure highlight this relationship between mechanical and chemical loadings.

The analyses of the main influence of each part are necessary in PWR environments. So Framatome ANP's Technical Center has adapted a specific tribometer to allow electrochemistry measurements. Three-electrode mounting permits us to regulate or to access current or potential under PWR conditions, with regard to the mechanical solicitation induced in the test apparatus. A first static test series showed the influence of temperature on the electrochemistry response and the good reproducibility of sequence in specific aqueous environments. This result revealed the importance of test conditions and the need to integrate corrosion kinematics.

Scratch tests do not lead to a significant depassivation–repassivation phenomenon in the environment considered. It is shown that the limited dimensions of the indent mark cannot provide a high current density. However, corrosion kinetics is clearly recognizable. Based on this last experiment, orbital configurations have been carried out to increase the contact area. The electrochemical phenomena are clearly identified in relation to mechanical parameters.

The future objectives are to carry out tests in several configurations and to estimate the relative importance of mechanical and corrosion processes.

9.5 References

1. D. Landolt, *Corrosion et chimie de surface de métaux, Traité des matériaux*, Presses Polytechniques et Universitaires Romandes, 1997.
2. D. Landolt, P. Jemmely, S. Mischler, Aspects électrochimiques et mécaniques du comportement tribologique des métaux passivables en milieux aqueux, *Actes JFTC 98*, 1998, 21–42.
3. E. Lemaire, M. Le Calvar, *Wear*, 2001, **249**, 338–344.
4. P. Ponthiaux, F. Wenger, J. Galland, P. Kubecka, L. Hyspecka, Interaction between friction and corrosion, *Actes JFTC 98*, 1998, 35–42.
5. P. Cohen, *Water Coolant Technology of Power Reactor*, American Electric Society, 1980.
6. D. Kaczorowski, J. M. Georges, S. Bes, A. Tonck, A. B. Vannes, J.-P. Vernot, *C. R. Acad. Sci. Paris*, 2001, t.2, Série IV, 739–747.
7. B. Stellwag, The mechanism of oxide film formation on austenic stainless steel in high temperature water, Corr. *Sci.*, 1998, **40** (2/3), 337–370.
8. R.-W. Bosch, B. Schepers, M. Vankeeberghen, Development of a scratch test in an autoclave for the measurement of repassivation kinetics of stainless steel in high temperature high pressure water, *Electrochim. Acta*, 2004, **49**, 3029–3038.

10

Tribocorrosion in pressurized water reactor environments: room temperature results and finite element modeling

D. DÉFORGE, A. AMBARD and A. LINA,
Electricité de France (EDF) R&D, France;
P. PONTHIAUX and F. WENGER, Ecole Centrale Paris, France

10.1 Introduction

In pressurized water reactor (PWR) nuclear power plants, many mechanical components suffer from wear damage due to flow-induced vibrations. A recent review of the wear scar patterns by Ko [1] shows that corrosion is often involved in wear phenomena. Moreover, tribocorrosion is proven to be active in wear of control rod mechanisms [2, 3]. Electricité de France (EDF) is thus investigating the role of corrosion in the wear process taking place in high temperature (300–320°C) pressurized water (155 bar).

For about 20 years, different research groups have developed electrochemical methods to study wear in corrosive environments. An excellent review was recently published by Landolt *et al.* [4]. Several tribometers able to work in corrosive electrolytes under electrochemical control have been developed [5], but to our knowledge, none of them can work at high temperature (above 300°C). EDF has thus added an electrochemical apparatus to a tribometer able to operate in high temperature water.

In this paper, the device is presented, and first experiments on the tribo-electrochemical behavior of an AISI 304L stainless steel are reported. All tests are done at room temperature for the purpose of comparison with a classical pin-on-disk tribometer.

10.2 Presentation of the Electricité de France tribometer

The machine (Fig. 10.1) was designed and manufactured by Atomic Energy of Canada Limited (AECL). It is an autoclave containing a cylindrical wear specimen and a cantilevered tube. The tube passes through a seal assembly to the outside of the autoclave. Its motion is generated by a vibration generator consisting of two-step motors connected in parallel and driven

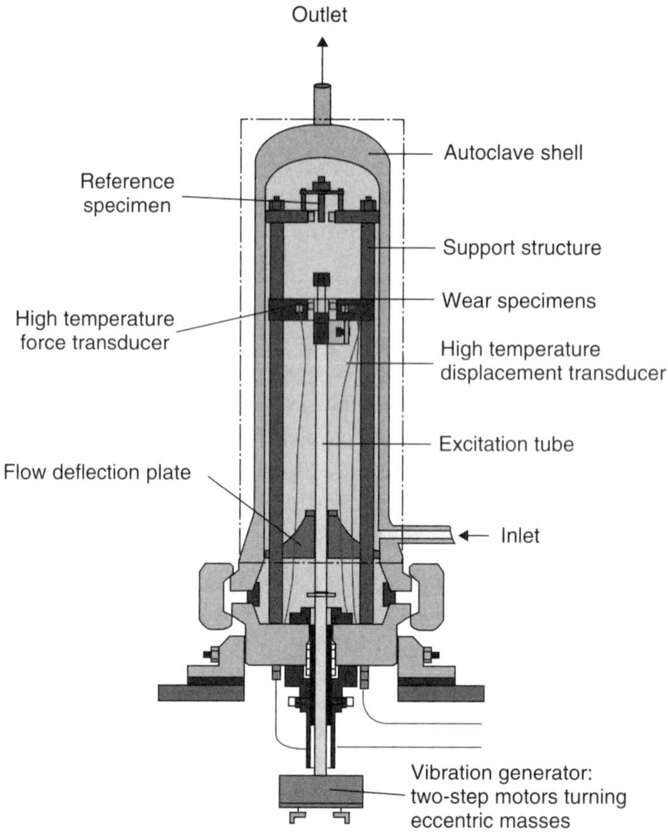

Fig. 10.1 The Electricité de France (EDF) wear simulator. A motor generates the vibration of a cylindrical tube inside a fixed cylindrical guide. Vibrations induce impact and sliding, but no rotation.

synchronously in opposite directions. Each motor drives an eccentric mass. Motor speed, mass values and their relative orientation can be varied to generate vibrations. Different motions likely to occur in service can be generated, from the easiest ones (pure impact/pure sliding) to the more complicated ones (combination of impact and sliding with different frequencies and locations). Displacement and force transducers able to operate at high temperature are used in order to record the contact conditions resulting from vibrations. The average normal load can be varied from 0 to 6N, and the vibration frequency between 14 and 30Hz. A more detailed description of the mechanical aspects of the device can be found elsewhere [6].

The original machine was modified to perform electrochemical measurements on the wear samples (working electrode). These samples – tube and

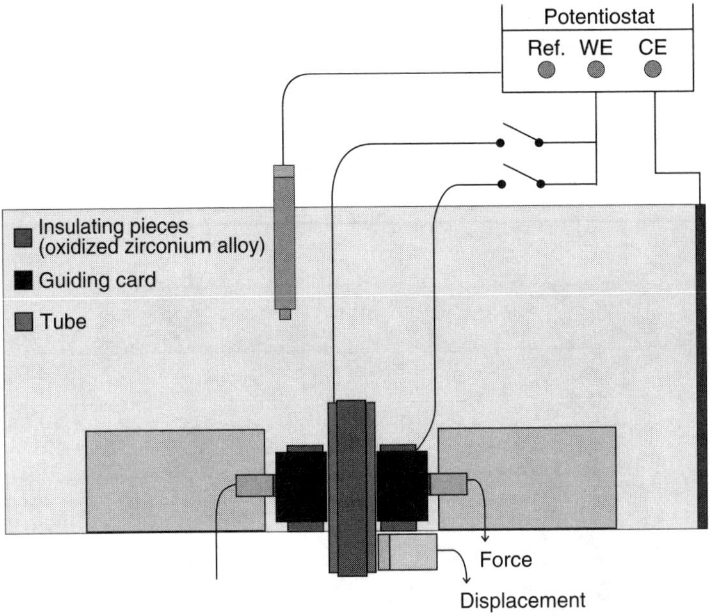

Fig. 10.2 Electrochemical connections of the wear simulator. The wear samples are insulated from the autoclave using zirconia pieces and connected as working electrodes of a three-electrode cell.

guide card – are insulated from the structure using high temperature resistant zirconia pieces. A hole in the autoclave shell enables insertion of a high temperature reference electrode, designed by EDF, a few centimeters away from the samples. The stainless steel autoclave is used as counter-electrode. The final schematic corrosion cell is shown Fig. 10.2. During the experiment, the tube and the guide can be short-circuited by connecting their electrical output outside the autoclave so that the same potential is always applied to both of them, even when they are not in contact. The resulting cell is quite far from ideal electrochemical conditions, especially due to the ohmic drop. Therefore, tests were performed at ambient temperature to allow comparison with results from another tribometer, and finite element modeling (FEM) simulations were done to get a better idea of the current repartition in the cell.

10.3 Experiments

10.3.1 Test material and test solutions

Measurements were performed using AISI 304L stainless steel (Table 10.1). Samples – tube and guiding card – are represented in Fig. 10.3. The test

Table 10.1 AISI 304L composition

	Element									
	C	Mn	Si	S	P	Ni	Cr	Mo	Cu	N_2
Weight fraction (%)	0.029	1.86	0.37	0.004	0.029	10.00	18.00	0.04	0.02	0.056

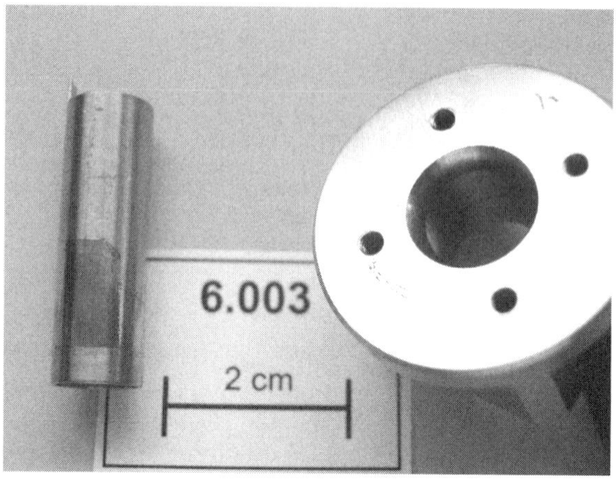

Fig. 10.3 Wear samples after a wear test and cleaning. On the left, the mobile tube which is plugged into the longer excitation tube. On the right, the guiding card which is fixed to the autoclave and electrically isolated from it.

solution was a mixture of boric acid (1000 ppm of boron) and lithia (130 ppm of lithium) dissolved in high purity distilled aerated water. This electrolyte had a low conductivity (210 μS cm^{-1}) and an initial pH of 8.7. Taking into account the modification of the water ionic product with temperature, this slightly basic pH was representative of the in-reactor pH at 320°C (the water ionic product decreases with temperature from 14 to 10.7 at 320°C, and the in-reactor pH lies around 7.1–7.3).

Mechanical conditions

A constant vibration frequency of 22 Hz was applied to the tube, leading to a pure orbital motion with the same frequency (Fig. 10.4). The average mechanical parameters resulting from the vibration are gathered in Table 10.2. Full contact between the tube and its guiding card was not always achieved. In order to take into account the contact discrepancies, a parameter was defined, the contact rate r, the percentage of time during which the normal load is above 0.2 N. The contact rate, and the actual

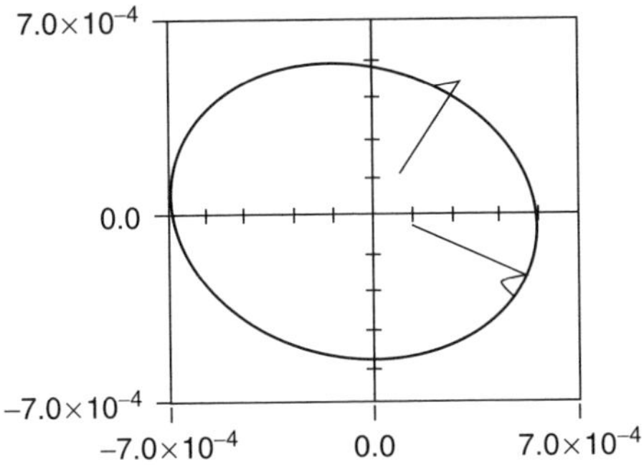

Fig. 10.4 A typical orbital motion of the tube inside its guide recorded during the wear test. Units: meter.

Table 10.2 Mechanical conditions of the wear test

	Average	Maximal	Standard deviation
Global load (N)	3.76	11.2	1.98
Normal Load (N)	3.00	9.93	1.94
Tangential load (N)	0.847	7.42	2.04

motion resulting from the vibration of the tube varied during the long term wear test, when the tube or the card became worn.

Electrochemical conditions

The electrochemical control was achieved using a PAR 263A potentiostat with floating ground. A reference electrode developed by EDF R&D for high temperature test conditions was used. It consisted of an Ag||AgCl cell (0.372 V vs standard hydrogen electrode at 20°C and atmospheric pressure) thermally isolated from the autoclave. All potentials were reported vs this reference. Classical electrochemical techniques were used to characterize the behavior of the passive film before and during sliding. Information on these electrochemical techniques for tribocorrosion studies has been given elsewhere [7]. Different kinds of measurements were applied:

- Recording the sample open circuit potential (OCP) when sliding started and during sliding.
- Potentiostatic control when sliding started and during sliding. A slightly anodic potential lying in the passive region of the steel, from 0 V, was chosen.

- Potentiodynamic curves recorded before sliding and during sliding. The potential scan rate was fixed at $1\,mV\,s^{-1}$.

Ex-situ wear analysis

The worn specimens were analysed to evaluate wear and wear pattern. The weight loss was measured using a ±10 μg microbalance and three-dimensional profilometry. The wear scars were examined using optical and scanning electron microscopy.

10.4 Results

10.4.1 Triboelectrochemical measurements: open circuit potential and potentiostatic study

Thirty minutes after immersion in the solution the OCP of the specimens stabilizes around –200 mV. When sliding starts, the OCP decreases and reaches a steady state potential, $E_M = -410\,mV$ (Fig. 10.5, top left). The potential recorded during the wear process is a mixed potential reflecting the state of both worn and unworn parts of the material. Abrasion causes local thinning or removal of the passive film, which makes the worn area more sensitive to corrosion. As the passive film regrows, a partial anodic current flows from the worn area which modifies the global electrochemical state of the sample. The OCP shifts toward less noble values where the oxygen reduction equilibrates the metal oxidation. This result is classical for passivated material in aerated water and is qualitatively reproduced using a pin-on-disk tribometer (Fig. 10.5, bottom left).

During a potentiostatic experiment at $E_M = -410\,mV$ without sliding, an average cathodic current of –30.2 μA is measured. As E_M is the OCP during sliding, the anodic shift of the global current due to the wear process is thus from –30.2 to 0 μA. By neglecting the variation of the oxygen reduction due to sliding at E_M, the anodic current may be taken as equal to the increase of the metal corrosion.

The corrosion enhancement in the wear track is also observed during potentiostatic experiments at $E = 0\,V$ (Fig. 10.5, right). This potential lies in the middle of the passive range, and very low current densities are measured (average current 0.80 μA). When sliding starts, the anodic current increases quickly to reach in a few seconds a steady state with an average current of 97.0 μA (standard deviation 8.18 μA). As explained concerning the open circuit potential, this increase is due to the growth of the passive film which is periodically removed. It is qualitatively in good accordance with the results obtained with a pin-on-disk tribometer (Fig. 10.5).

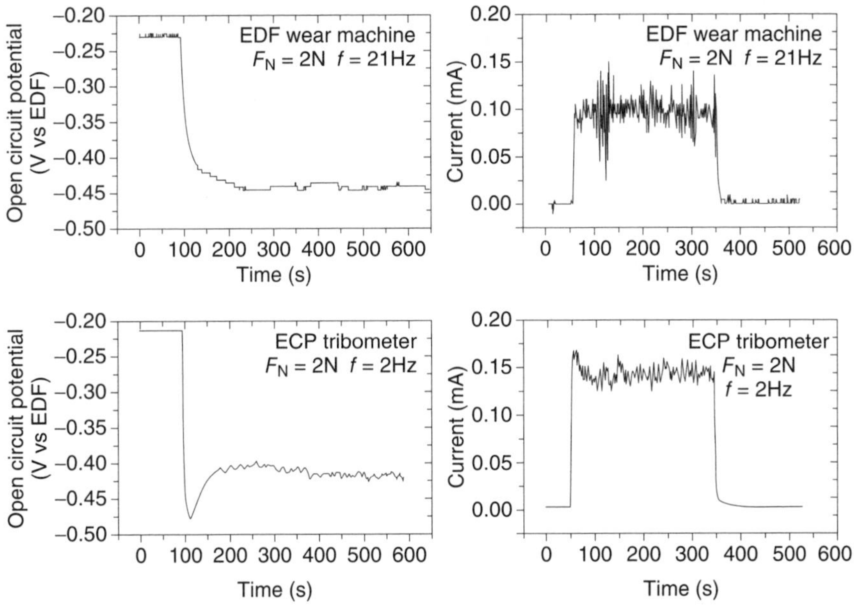

Fig. 10.5 Triboelectrochemical characteristics of the AISI 304L. First line: Electricité de France (EDF) tribometer. Second line: pin-on-disk tribometer. Left: open circuit potential versus time (sliding starts at $t = 100\,\text{s}$). Right: current evolution during sliding under potentiostatic conditions ($E = 0\,\text{V}$, sliding starts at $t = 50\,\text{s}$ and stops at $t = 250\,\text{s}$). ECP, electrochemical corrosion potential.

In a first approximation, it is possible to use the variation of the anodic current observed at $E_M = -410\,\text{mV}$ and $E = 0\,\text{V}$ to compute corrosion rates using Faraday's law:

$$\dot{m}(t) = \frac{i(t) \cdot M}{n \cdot F} \qquad [10.1]$$

where $\dot{m}(t)$ is the mass of material corroded per unit time (for example in milligrams per month), M the stainless steel average molecular weight, F the Faraday's constant (which is the charge of one mole of electrons) and n the average valence of dissolution of metal taken as the sum of the valence of the elements of the alloy ponderated by their atomic fraction. Using the numerical values $n = 2.628$, $M = 55.07\,\text{g}\,\text{mol}^{-1}$ and $F = 96\,490\,\text{C}\,\text{mol}^{-1}$, one finds a weight of stainless steel oxidized per unit charge $M/n \cdot F = 0.217\,\text{g}\,\text{C}^{-1}$. This value enables computation of the corrosion rates given in Table 10.3. The value is higher at $E = 0\,\text{V}$ than at E_M which may correspond to the rebuilding of a thicker passive film under anodic polarization.

Table 10.3 Current variations Δi during the short-time sliding test for both electrochemical conditions. i_{ws}: current without sliding; i_s: current under sliding; $\Delta m/\Delta t$: corrosion rate

	i_{ws} (μA)	i_s(μA)	Δi(μA)	$\Delta m/\Delta t$ (mg month^{-1})
$E_M = -0.410$ V	−30.2	0	30.2	17.2
$E = 0$ V	0.80	97.0	96.2	54.9

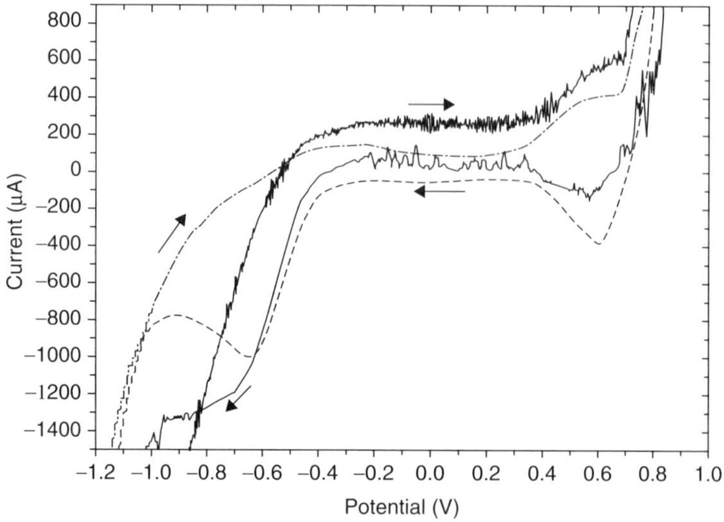

Fig. 10.6 Polarization curves of the AISI 304L in the boric acid/lithia media. Surface area = 40 cm^2, potential scan from −1.4 to 1.1 V, scan rate 1 mV s^{-1}. Dashed curves were recorded before the wear test and the solid curve during the wear test.

Potentiodynamic curves of the AISI 304L

Polarization curves of the as-received AISI 304L recorded before the wear experiment are shown in Fig. 10.6 (dashed curves). After cathodic polarization at −1.5 V, a small anodic peak is observed around −0.5 V, which may be due either to the regrowth of a passive film or to the oxidation of adsorbed hydrogen. The passive range extends up to 0.3 V. Above this potential, the current increases to a secondary passivation state. An *E*–pH diagram of chromium at ambient temperature indicates that this may be due to the oxidation of Cr^{3+} to Cr^{6+}. The oxidation of water is observed above 0.8 V, which leads to oxygen production. On reverse potential scanning, reduction of Cr^{6+} is observed. There is no active region below, and the processes at work are the reduction of dissolved oxygen below −0.35 V, and the reduction of water protons below −1 V, leading to hydrogen production.

Polarization curves recorded when vibrations are applied are also shown Fig. 10.6 (solid curves). Currents are higher during the wear experiment than on the as-received material between –0.35 and 0.8 V. No electrochemical reactions involving oxygen occur in this range of potentials. Since direct scans follow a cathodic polarization which produces hydrogen, it is safer to compare currents on reverse scannings than on direct scannings: hydrogen oxidation hardly influences their values. The current increase can then be attributed to the metal corrosion (dissolution and/or oxide regrowth). Between –0.35 and 0.3 V, the average current increase is 99.7 ± 32.6 μA. This is nearly equal to the value reported at $E = 0$ V.

Wear test

Wear experiments are performed for a period of 6 days at OCP. Variations of its value are observed during the test (amplitude 50 mV), which are correlated with the motion discrepancies, as shown by comparing the OCP and the contact rate (Fig. 10.7). It shows that the primary factor for depassivation is contact, and that a load above 0.2 N is enough to induce depassivation. The higher the contact rate, the greater is the depassivated area, and the lower the potential. This is in total accordance with the interpretation given in the previous sections.

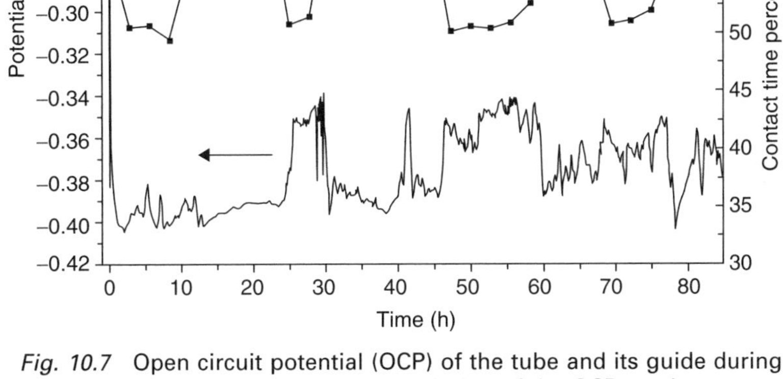

Fig. 10.7 Open circuit potential (OCP) of the tube and its guide during a long-time wear experiment. The evolution of the OCP can be connected to the evolution of the contact rate (percentage of time for which $F_N > 0.2$ N).

Fig. 10.8 Scanning electron micrograph of the worn tube. The top part of the photograph corresponds to some oxide deposited on the border of the worn area.

A typical wear scar is shown in Fig. 10.8. After 6 days, reddish oxides are present all around the worn area. Most of the particles are deposited on the card. The worn area looks like having been mechanically polished and almost no oxides are observed on it. The oxide deposit has a quite low adherence and can be removed using an ethanol bath. The underneath surface is not attacked and looks as-received. The oxide localization around the wear track may be due either to its ejection from the contact area by the tube or to the fact that precipitation of the oxide occurs at this location, where the local pH may be higher due to oxygen reduction [8]. Weight losses of the tests are given in Table 10.4. By extrapolating the results from the electrochemical measurements, it seems that the volume removed by oxidation (whether it is due to anodic dissolution or film regrowth) may account for one-quarter of the global weight loss. This division of wear between a mechanical and a corrosion part is often done in the literature, and results usually vary between 0 and 100% [9]. The 'mechanical part' of wear, deduced from the electrochemical part, is due to the occurence of some adhesion or abrasion processes, which can lead to the ejection of mechanical particles before they become oxidized. For example (Fig. 10.8), abrasive marks are present along the wear track, elongated in the direction of motion. This approach, consisting of dividing the wear resulting from corrosion and mechanical removal, is dangerous as corrosion may modify deeply the tribological conditions in the wear track: corrosion products can form a third body with abrasive properties, leading, for a given mechanical solicitation, to a much higher 'mechanical part' than in a noncorrosive environment.

Table 10.4 Weight variations during a 6 day wear test performed at open circuit potential

	Weight loss (mg) of test ref. 6.003	Equivalent in mg month^{-1}
Card	34.3	174.0
Tube	7.17	36.4
Total	41.47	210.4

The relative importance of the corrosion processes in the weight loss and the presence of solid corrosion products in the wear track prove that it is important to develop a global approach to wear taking into account both mechanical and electrochemical parameters.

10.5 Finite element modeling of the electrochemical results

10.5.1 Background

During a wear experiment, the sample surface is highly heterogeneous and always disturbed due to the pin motion. The current-line distribution is thus permanently disturbed and results from the interaction between the unworn area (with high electrochemical impedance), the active area created by the tube vibration (with low impedance) and the area which repassivates itself after a previous contact (with an intermediate impedance).

In a first approximation, it is possible to simplify the problem by considering that the global surface A_T is divided in two parts with distinct electrochemical characteristics: an active area A_A and a passive area A_P. When the steady state is reached, the fraction of active area may be considered constant.

The current repartition of the specimens depends on the experiment. If j_A and j_P are the current density flowing respectively from the active and passive areas, they verify:

$$A_A \cdot j_A + A_P \cdot j_P = 0 \quad [10.2]$$

for open circuit conditions, and

$$A_A \cdot j_A + A_P \cdot j_P = I \quad [10.3]$$

when a potential is applied, where I is the measured current.

j_A and j_P depend on the intrinsic characteristics of the interface, especially its potential, Φ. In a first approximation, a linear relationship between current and potential can be considered, which can be written:

$$j_i = \frac{\Phi - E_i}{A_i \cdot R_i}; \quad i = A, P \qquad [10.4]$$

E_A and E_P are the intrinsic equilibrium potentials of the active/passive interface, traducing their relative nobility; and R_A and R_P are the polarization resistance of the active/passive interface, traducing their ability to become polarized. This simple model explains schematically the electrochemical behavior of the sample. For example, the potential drop can be considered as a mixed potential resulting from a galvanic coupling between both areas (Fig. 10.9). The bare surface potential tends to be lower than that of the passive surface, and the potential observed during sliding is lower, with a value derived from Eq. 10.2 and 10.4:

$$\Phi = E_M = \frac{R_A \cdot E_P + R_P \cdot E_A}{R_A + R_P}. \qquad [10.5]$$

Even if the total current flowing out of the surface is zero (Eq. 10.2), there is an exchange current between active and passive areas. The active area is polarized anodically, which increases the repassivation kinetics, and the passive area is polarized cathodically.

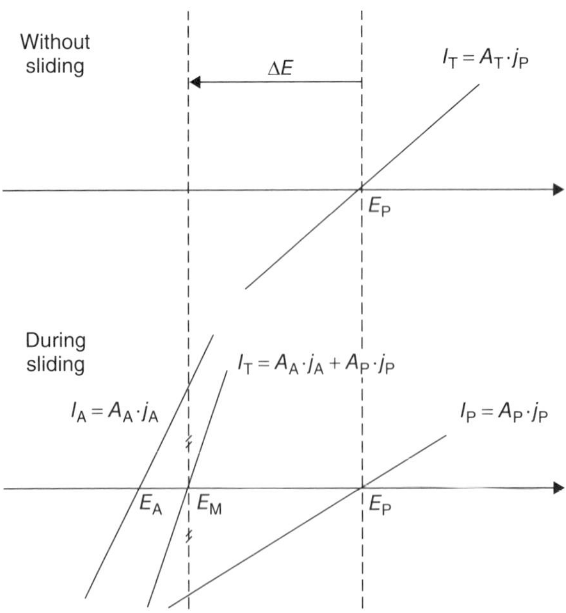

Fig. 10.9 Schematic explanation of the potential drop due to sliding using the formalism developed in Section 10.5.1: the open circuit potential observed during the wear experiment. E_M is a mixed potential between E_A and E_P. I_T: total current.

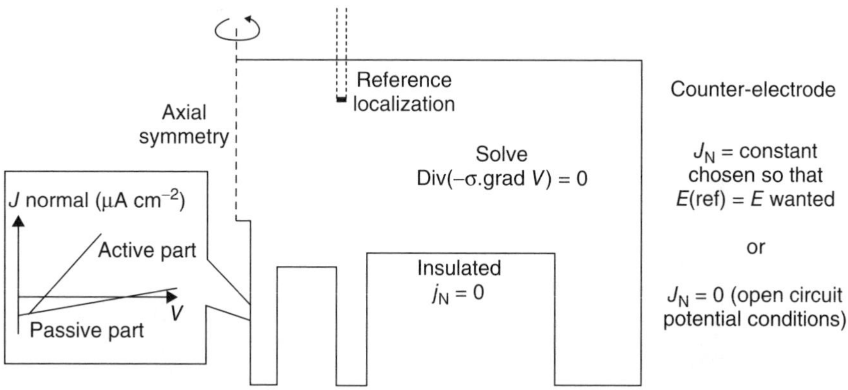

Fig. 10.10 Schematic explanation of the finite element modeling of the cell. The simulation consists of a resolution of $\mathrm{div}(-\sigma.\vec{\mathrm{grad}}(V)) = 0$ in the solution using the boundary conditions reported.

An idea of the current distribution between these parts can be obtained using finite element modeling. The commercial software FEMLAB 3.0 is used. (FEMLAB is a trademark from COMSOL.) For the sake of simplicity, the cell is considered to have an axial symmetry (Fig. 10.10), and the active area is assumed to be uniformly distributed on the whole sample circumference (this hypothesis is possible due to the fact that the rotation frequency of 20 Hz is high with respect to the time necessary for the specimen to repassivate). The simulation consists of a resolution of the equation: $\Delta\Phi = \mathrm{div}(-\sigma \cdot \vec{\mathrm{grad}}(V)) = 0$, where Φ is the potential. This equation establishes the local conservation of the current. The conductivity of the water is taken as equal to $210\,\mu\mathrm{S\,cm^{-1}}$. The cell is discretized. Boundary conditions are imposed. They depend on whether a potential is applied or not.

10.5.2 OCP experiment

For open circuit conditions, all borders are considered as insulating except the sample surface. For the passive area, Eq. 10.4 is applied, with an intrinsic potential E_P of −200 mV and a polarization resistance R_P of 10 kΩ. The value of the potential is derived from measurements performed before sliding, when the whole sample is passive, and the value of the polarization resistance is computed in an independent experiment reported elsewhere [10].

The same kind of relation is used for the active area, but it is harder to access the parameters R_A and E_A, due to the fact that the naked material is unstable in the solution. A value of −1.6 V is chosen for E_A, which is the lowest OCP reached after cathodic depassivation, and the slope R_A is chosen as a fitting parameter: R_A is adjusted so that the potential read at

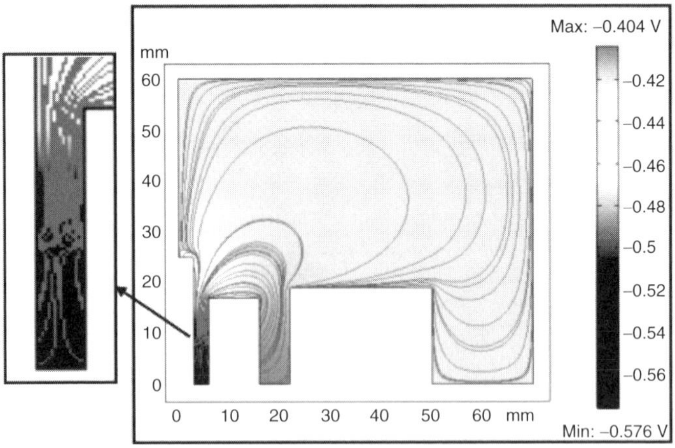

Fig. 10.11 Finite element model of the cell for open circuit conditions. Surface: electric potential; streamline: total current density. The sample's average potential $E_S = -410$ mV. The actual potential of the samples lies between −410 and −576 mV.

the point where the reference electrode is located corresponds to the experimental OCP, i.e., −410 mV.

Graphical results are displayed in Fig. 10.11. It can be observed that the potential is not homogeneous over the whole cell. Current lines flow from the active area to the passive area, and the chosen parameters lead to a potential distributed between −0.410 and −0.576 V on the surface of the specimen. The software enables us to compute the value of the current flowing out of the active part. It gives a value I_A which is of the same order of magnitude as the experimental value reported for open circuit conditions (Table 10.3):

$$I_A = \iint_{A_A} j_A \cdot dA_A = \left| \iint_{A_P} j_P \cdot dA_P \right| \approx 23\,\mu\text{A}. \qquad [10.6]$$

10.5.3 Potentiostatic experiment

Potentiostatic experiments are modeled assuming the same behavior as in the previous section with the difference that the counter-electrode is considered as a homogeneous current source (primary current distribution). The value of the current flowing out is adjusted so that the potential near the reference electrode is the imposed potential 0 V: precisely how a potentiostat proceeds to adjust a potential. The resulting distribution is displayed in Fig. 10.12.

The potential distribution at the surface of the specimen is heterogeneous. It varies between 0 V, the imposed value, and −0.305 V in the wear track.

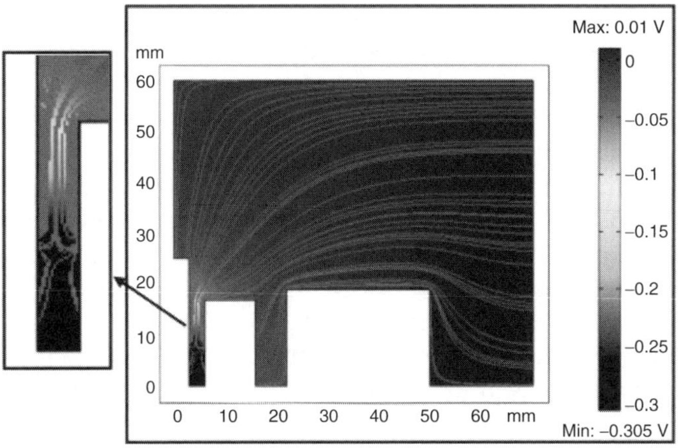

Fig. 10.12 Finite element model of the cell for potentiostatic conditions at $E = 0\,V$. Surface: electric potential; streamline: total current density. The sample potential actually lies between 0 V and –305 mV. Part of the current flowing out of the active area is not sensed by the counter-electrode.

The difference between applied and 'real' potentials is particularly high in the small area between the two samples: the diameter clearance between both samples is $d = 0.6\,mm$. The current lines show that in the immediate neighborhood of the sample, the passive area is polarized cathodically (below E_P), and thus provides a cathodic contribution to the global equilibrium (which may correspond to some oxygen reduction). Some current lines begin and terminate on the surface of the sample and do not reach the counter-electrode. The current measured at the counter-electrode does not take into account this galvanic effect. Some current flowing to the counter-electrode also comes from the passive part of the sample. The measured current thus results from the interaction between these different phenomena, and it may differ significantly from the oxidation current in the wear track. According to the simulation, the global current flowing out of the active part is finally in reasonable accordance with experimental results (Table 10.3):

$$I_A = \iint_{A_A} j_A \cdot dA_A = 70\,\mu A. \qquad [10.7]$$

10.6 Conclusions and perspectives

This preliminary study shows that under mechanical conditions representative of those encountered in primary water, the impact-sliding motion degrades the passive film formed on AISI 304L stainless steel at room temperature. This degradation leads to a huge increase of corrosion in the wear track, and to the formation of some oxide deposits. Final oxidation rates are important in regard to the final wear rate. Any attempt to model wear in this environment should thus take corrosion into account. Mechanical and electrochemical processes have to be considered as a whole, leading to the material degradation.

Electrochemical measurements are also interesting as they provide in-situ data enabling characterization of the intensity of the wear process: in-line OCP measurements enable determination of whether a given vibration leads to the degradation of the passive film or not, and the more wearing the motion, the lower is the resulting OCP.

Similar experiments will now be performed in water at high temperature (320°C). More work will also be performed with FEM simulation: non-linearity between current and potential will be introduced to increase the representativeness of the results.

10.7 Acknowledgments

T. Ghys is acknowledged for his technical support and his great help in performing the experiments. This work is performed under the auspices of EDF R&D program, grant P001N0.

10.8 References

1. Ko, P. L., Wear of power plant components due to impact and sliding. *Appl. Mech. Rev.*, 1997, **50**(7), 387–411.
2. Lemaire, E. and M. Le Calvar, Evidence of tribocorrosion wear in pressurized water reactor. *Wear*, 2001, **249**, 338–344.
3. Benea, L., P. Ponthiaux, F. Wenger, D. Hertz and J. Y. Malo, Tribocorrosion of stellite 6 in sulphuric acid medium: electrochemical behavior and wear. *Wear*, 2004, **256**(9–10), 948–953.
4. Landolt, D., S. Mischler and M. Stemp, Electrochemical methods in tribocorrosion: a critical appraisal. *Electrochem. Acta*, 2001, **46**, 3913–3929.
5. Mishler, S. and P. Ponthiaux, A Round Robin on combined electrochemical and friction tests on alumina/stainless steel contacts in sulphuric acid. *Wear*, 2001, **248**, 211–225.
6. Lina, A. *et al.*, The influence of water flow on the impact/sliding wear and oxidation of PWR control rods specimens. *Wear*, 2001, **251**, 839–852.
7. Ponthiaux, P., F. Wenger, D. Drees and J.-P. Celis, Electrochemical techniques for studying tribocorrosion processes. *Wear*, 2004, **256**(5), 948–953.
8. Evans, U. R., *The Corrosion and Oxidation of Metals*, Arnold, London, 1960.
9. Mischler, S., S. Debaud and D. Landolt, Wear-accelerated corrosion of passive metals in tribocorrosion systems. *J. Electrochem. Soc.*, 1998, **145**(3), 750–758.
10. Déforge, D., F. Huet, R. P. Nogueira, P. Ponthiaux and F. Wenger, Electrochemical noise analysis of tribocorrosion processes under steady-state friction regime. *Corrosion 2005*, Houston, No. 05369.

11

Tribocorrosion of stellite 6 alloy: mechanism of electrochemical reactions

FRANÇOIS WENGER and PIERRE PONTHIAUX, Ecole Centrale Paris, France;
LIDIA BENEA, Dunarea de Jos University of Galati, Romania;
JEAN PEYBERNÈS, Commissariat à l'Energie Atomique (CEA), France; ANTOINE AMBARD, Electricité de France (EDF), France

11.1 Introduction

In pressurized water reactors (PWRs) of nuclear power plants, the gripper latch arms of the control rods' command mechanisms are protected from corrosion and wear by thick stellite 6 layers. A few years ago, it was suggested [1] by damage surveys of numerous latch arms worn in PWRs, that tribocorrosion was involved in the degradation of these layers. Tribocorrosion is the process leading to wear by a combined action of friction and corrosion.

This conclusion was drawn from the following observation: the wear loss W of the gripper latch arms not only increases with the number of steps N (mechanical interaction between an arm and the control rod, involving impact plus sliding friction) but also increases, for a given number of steps, with the average duration between two successive steps (latency time τ). An empirical law was proposed:

$$W = Nw_0\left(\frac{\tau}{t_0}\right)^{(1-n)} \qquad [11.1]$$

where n is a dimensionless parameter, with value close to 0.6, and w_0 is the average mass loss per step, while t_0 is a time constant.

This law was explained by a tribocorrosion mechanism in which the wear of the material is mainly due to the periodic mechanical destruction (during friction step) of the passive oxide film (Cr_2O_3 film) protecting the alloy from the aqueous environment, and restoration of this film by oxidation of the bare alloy areas during the subsequent latency period.

With this mechanism, the evolution $i(t)$ of the oxidation current on the bare metal areas during the latency periods can be derived from Eq. 11.1 [2]:

$$i(t) = i_0 \left(\frac{t}{t_0} \right)^{-n}. \qquad [11.2]$$

When a tribocorrosion process is involved, the wear is often expressed as the sum of an electrochemical component, W_e, and a mechanical one, W_m:

$$W = W_e + W_m. \qquad [11.3]$$

In this case the total wear W can be identified with the component W_e since the wear rate is controlled by the kinetics of the oxidation reaction leading to passive film growth. The component W_m corresponds to the mass of bare alloy removed by a purely mechanical action. In this case, W_m seems to be much smaller than W_e.

In a previous study [3], the tribocorrosion of stellite 6 was studied in the laboratory, at ambient pressure in a solution of sulfuric acid (0.5 M) at 20°C. In this solution, which is much more oxidizing than the pressurized water medium, accelerated corrosion and tribocorrosion tests could be carried out. A pin-on-disc tribometer was used and a test procedure of intermittent sliding friction was applied in cycles consisting of a short step of friction, followed by a longer latency step without applied friction. The same type of wear law (Eq. 11.1) was found, suggesting that similar tribocorrosion mechanisms are active in sulfuric acid as in a PWR environment.

Therefore, a more detailed investigation of the tribocorrosion mechanism was undertaken. The study was carried out at ambient pressure in two electrolytes:

1 0.5 M sulfuric acid at 20°C.
2 A solution of boric acid (1000 ppm of boron) and lithia (12 ppm of lithium) at 85°C. This solution was chosen because its chemical composition and properties are close to those of water in PWRs.

The following tests were implemented:

1 Tribocorrosion tests in conditions of continuous or intermittent friction with wear measurements.
2 Potentiokinetic polarization curves.
3 Analysis of the transient current responses of the alloy to anodic potential jumps of different amplitudes.

11.2 Experimental conditions

The specimens for tribocorrosion tests were prepared by plasma spraying of thick layers (1 mm) of stellite 6 at one end of stainless steel (304 L) cylinders

Table 11.1 Composition of the stellite 6 layer

	Element						
	O	Si	Cr	Fe	Co	Ni	W
Weight %	6.4	2.1	28.3	3.0	55.4	2.4	2.4

of 25 mm diameter and 25 mm height. The composition of stellite 6 is given in Table 11.1.

In the tests in water or water + B + Li, the working surface was the disc-shaped end of the stainless steel cylinders, covered by stellite 6. The area of the disc is 4.91 cm^2.

In the tribocorrosion tests carried out in sulfuric acid to determine the wear law, the disc-shaped end of the cylinder was partly 'masked' by an insulating cataphoretic paint, and the working surface was reduced to a ring (internal diameter 14 mm; external diameter 18 mm) of 1.0 cm^2 area. This delimitation of the working surface was used in our previous study of stellite 6 in sulfuric acid to prevent depassivation and corrosion on the areas along both sides of the wear track [3].

Before each test, the working surface was polished with emery paper (grade 1200) and rinsed with distilled water.

For tribocorrosion pin-on-disc tests, the specimens were mounted in a cell, containing the electrolyte and electrodes, with the working surface of the specimens facing upwards. The counterbody (pin) was a corrundum cylinder with a flat end (2.0 mm diameter; apparent contact area = 3.14 mm^2), mounted vertically on a rotating head, on top of the specimen. The lower end of the pin was applied against the stellite disc at an adjustable normal force F_n in plane-on-plane contact conditions. When rotation was applied, the end of the pin drew a circular wear track (16 mm in average diameter) on the working surface (see Fig. 11.1).

Tribocorrosion tests were carried out at average contact pressures of 120 MPa ($F_n = 377$ N) and 30 MPa ($F_n = 94$ N).

Both continuous and intermittent friction tests were carried out to obtain the experimental relationship between wear and latency time. In the intermittent tests, friction was applied periodically: during each cycle, friction was first applied for 2 s at an average contact pressure of 120 or 30 MPa at 120 r.p.m. and then stopped during a latency time of 20 or 200 s. This mechanical solicitation was repeated for 2500 cycles (10 000 rotations in all). Some features of these tests reproduce the wear conditions of stellite in the PWR: intermittent friction, average contact pressure of 120 MPa. Friction tests under continuous sliding conditions were carried out under the same tribological conditions (30 or 120 MPa; 120 r.p.m.; 10 000 rotations).

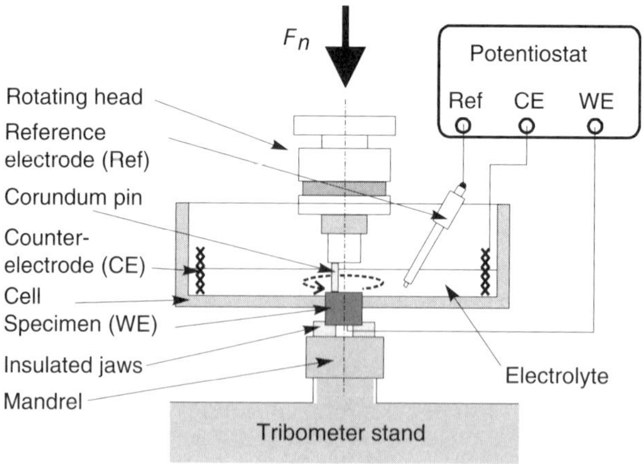

Fig. 11.1 Schematic representation of the electrochemical cell and of the mounting of the specimen on the tribometer.

Three aqueous solutions were used:

1 0.5 M H_2SO_4 prepared from pure sulfuric acid and distilled water. The tests were carried out in this electrolyte at 20°C.
2 A solution of boric acid and lithia, named 'Water + B + Li', containing 1000 ppm of boron and 12 ppm of lithium, with a pH of 7.5. The solution was prepared from pure boric acid and lithia dissolved in distilled water. The tests were carried out in this electrolyte at 85°C.
3 Additional tests were also carried out in distilled water at 20°C, in order to reduce the contribution of corrosion to a minimum and evaluate the mechanical component W_m of the total wear.

For electrochemical measurements, a three-electrode set-up was used, with the specimen as working electrode and a circular platinized titanium grid as counter-electrode. As reference electrode, an Hg/Hg_2SO_4/saturated K_2SO_4 solution electrode [SSE; $E = +670$ mV/standard hydrogen electrode (SHE)] was used in sulfuric acid at 20°C. In the boric acid and lithia solution at 85°C, an Ag/AgCl/saturated KCl solution electrode (Ag-AgCl; at 85°C, $E = +360$ mV/SHE) was taken. The electrodes were connected to a PAR 273A potentiostat and Corrware 2.2 (Scribner) software was used.

Polarization curves were recorded without friction and under continuous friction conditions, by applying a linear time-dependent potential scan (1.67 mV s^{-1}) from the cathodic hydrogen evolution potential domain up to the beginning of the anodic transpassive dissolution domain. The ohmic drop due to the resistance of the electrolyte was measured and corrected.

In the potential jump experiments, the potential steps were applied by the potentiostat and the current transients were recorded by a data acquisition chain (CBO Instrumentation) at an acquisition frequency of 4kHz.

At the end of the tribocorrosion tests, the wear volume was deduced from three-dimensional surveys of the wear track recorded with a high resolution optical microtopograph, at a lateral resolution of 1 μm and vertical resolution of 30nm. The volume of the wear track was measured and the corresponding weight loss was calculated. Direct weight loss measurements (accuracy: 0.2mg) were also carried out and compared to the values deduced from microtopographic measurements, in order to detect the possible development of corrosion out of the wear track.

11.3 Wear laws

The determination of the wear laws $W_e(\tau)$ was performed through tribocorrosion tests under continuous and intermittent sliding. It must be noted that, when continuous sliding is applied at 120r.p.m., every part of the wear track is in contact with the pin every 0.5s. Consequently, such a test can be considered as an intermittent test with a latency time of 0.5s. Consequently, the wear laws were determined from results of wear measurements corresponding to three values of the latency time, namely 0.5, 20 and 200s.

To calculate the electrochemical component W_e of the total wear, the mechanical component W_m had to be substracted from the measured wear W. The value of W_m was estimated from tribocorrosion tests carried out in distilled water. In this environment, the contribution of the oxidation processes to the total wear is expected to be much smaller than in the other electrolytes. Actually, the smallest wear was obtained in distilled water, and the variation of wear with latency time was found to be much smaller than in the other electrolytes (Fig. 11.2).

The oxidation reaction seems to be very slow in distilled water. During continuous sliding tests, the amount of oxide formed during the short latency period (0.5s) and then removed by the pin, should be smaller than the amount of bare material removed by a mechanical action of the pin. Hence, the contribution of the electrochemical component W_e to the total wear was assumed to be negligible and W_m was derived from the wear value measured after a test carried out in continuous sliding motion conditions:

$$W_m = (0.8 \pm 0.3)\ \text{mg}. \qquad [11.4]$$

This value of W_m was assumed to represent the mechanical component of wear, independent of the latency time, obtained after 10000 rotations, for the tribocorrosion tests carried out under a contact pressure of 120MPa in the other electrolytes. It must be noted that, for a contact pressure of 30MPa, the wear was too small to be measured with sufficient accuracy

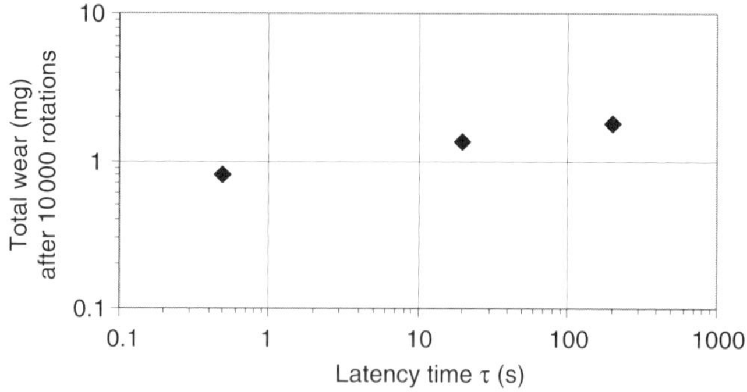

Fig. 11.2 Variation of the total wear of stellite 6 with latency time τ in distilled water at a contact pressure of 120 MPa.

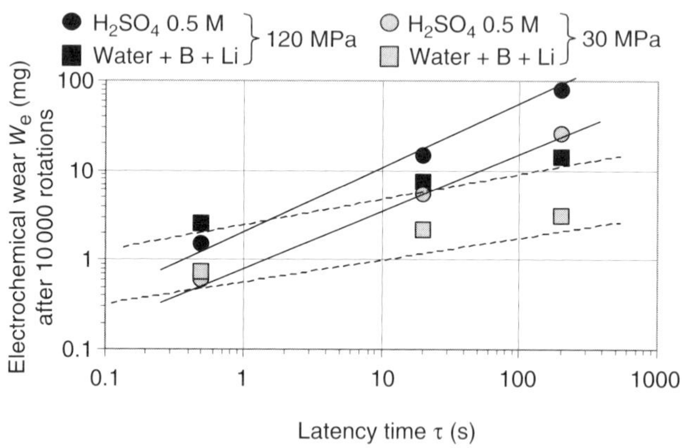

Fig. 11.3 Variation of the electrochemical component of wear W_e versus the latency time τ, for both electrolytes and two contact pressures.

($W_m \approx 0.2$ mg). In Fig. 11.3, the variation of W_e is plotted versus τ. The variations are in agreement with Eq. 11.1. The values of w_0, t_0 and n are gathered in Table 11.2.

It must be noted that the lines obtained at 120 and 30 MPa are parallel for a given electrolyte. This indicates that oxidation reaction kinetics involved in the repassivation of the surface after friction are the same at both contact pressures. The shift between the lines at 120 and 30 MPa can be explained by the difference in the real contact areas. When the apparent contact pressure is multiplied by 4 (from 30 to 120 MPa), the contact area increases by a factor of 1.9 for the sulfuric acid solution and 3.9 for the solution of boric acid and lithia. It must be noted that these factors do not give the ratios between the real contact areas, but the ratios between the

Table 11.2 Values of the parameters w_0, t_0 and n of Eq. 11.1 corresponding to the lines of Fig. 11.3

Electrolyte	P_{avg} (MPa)	w_0 (mg)	t_0 (s)	n
0.5 M H_2SO_4	30	1.6	1	0.40 ± 0.05
	120	3	1	0.40 ± 0.05
Water + B + Li	30	0.9	1	0.70 ± 0.05
	120	3.5	1	0.70 ± 0.05

areas depassivated by friction. Observations of the surface at the microscopic scale after tribocorrosion tests show that, in the wear track, depassivation and dissolution of the material do not take place only in the contact areas but also between these areas. As a result, the ratios of the values of w_0 are not identical to the ratios of real contact areas.

The value of n seems to be related to the oxidizing and corrosive properties of the electrolyte. If the value of n is small, the decrease of the oxidation current (Eq. 11.2) is slow. It could mean that the surface film can grow during a long period of time and that a large amount of metal can be oxidized during the latency period. This is the case in 0.5 M H_2SO_4. It must be noted that the growth of the passive film is perhaps not the only origin of the anodic current, and that parallel dissolution of the alloy might also occur, as shown by other authors [4] in the study of passivation of chromium and stainless steels in sulfuric acid. In addition, in our previous study of the wear law of stellite 6 in sulfuric acid by intermittent tribocorrosion tests [3], the occurrence and the influence of dissolution were pointed out. However, on areas where the film can grow in applied potential conditions, the dissolution of the metal should stop very quickly.

In water + B + Li, the value of n is higher, indicating that a thinner oxide film is removed at every friction step. The growth of the passive film seems to be slower. However, even though the growth of the film is slow, the dissolution of the alloy can remain negligible, if the rate of the dissolution reaction is smaller than the rate of passivation.

Finally, it is important to compare the values of n found here for the two electrolytes, with the values determined on the latch arms worn in a PWR environment [1]: $n = 0.6$. This value of n seems to rank the corrosiveness of the PWR environment under tribocorrosion solicitation between those of the H_2SO_4 solution (smaller value of n, higher corrosiveness) and of the solution of boric acid and lithia (higher value of n, lower corrosiveness).

Nevertheless, tribocorrosion tests under continuous and intermittent sliding conditions are not well suited for a detailed investigation of the kinetics of the oxidation processes, mainly for the following reason: the tests

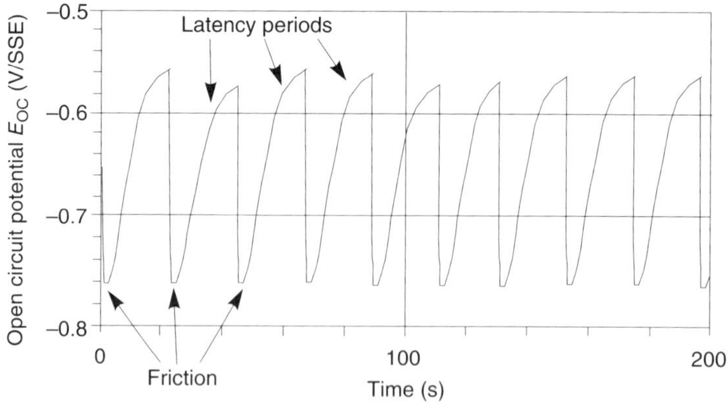

Fig. 11.4 Variations of the open circuit potential of stellite 6, during a tribocorrosion test with intermittent friction, in 0.5 M H_2SO_4. Latency time $\tau = 20$ s. SSE, Hg/H_2SO_4/saturated K_2SO_4 solution electrode.

are performed in open circuit (or free) potential conditions. During the test, the potential of the specimen undergoes periodic variations with large amplitudes (Fig. 11.4).

Under open circuit potential conditions, there is a galvanic coupling between the depassivated areas of the wear track and the surrounding areas remaining in a passive state. The instantaneous value of the potential depends on the ratio between depassivated and passive metal areas. It depends also on the thickness of the passive film in these areas. These parameters are continuously varying during an intermittent tribocorrosion test. Conversely, the kinetics of oxidation of the alloy depend on the potential, and so the passivation and dissolution (if possible) rates vary during the test.

To obtain more detailed information on the oxidation process of the alloy, electrochemical methods such as polarization curves and potential jumps can be used.

11.4 Electrochemical behavior of stellite 6

11.4.1 Polarization curves

The potentiokinetic polarization curves $I(E)$ of stellite 6 in water + B + Li and in the 0.5 M H_2SO_4 solutions are given in Fig. 11.5 and 11.6. The shape of the curves recorded in both electrolytes without applied friction reveals the existence of a wide passivation plateau (0.8 V) limited at the cathodic end by the hydrogen evolution and oxygen reduction domains, and at the

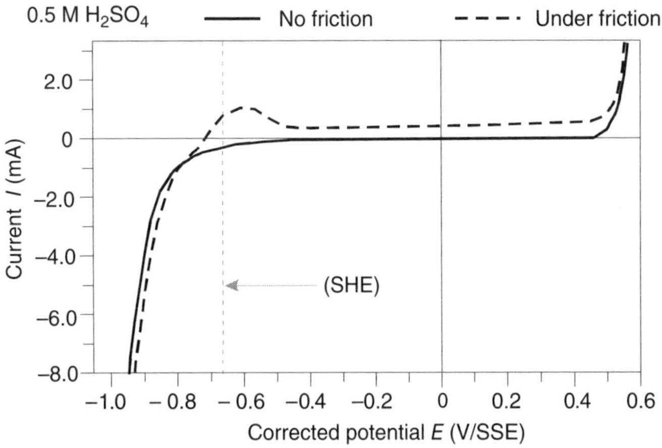

Fig. 11.5 Potentiokinetic polarization curves (corrected from ohmic drop due to the electrolyte resistance) of stellite 6 in 0.5 M H_2SO_4, with no friction and under continuous sliding friction (120 MPa; 120 r.p.m.). Working surface: disc (4.9 cm^2). The potential of the standard hydrogen electrode (SHE) (−0.67 V/SSE) is marked. SSE, Hg/H_2SO_4/saturated K_2SO_4 solution electrode.

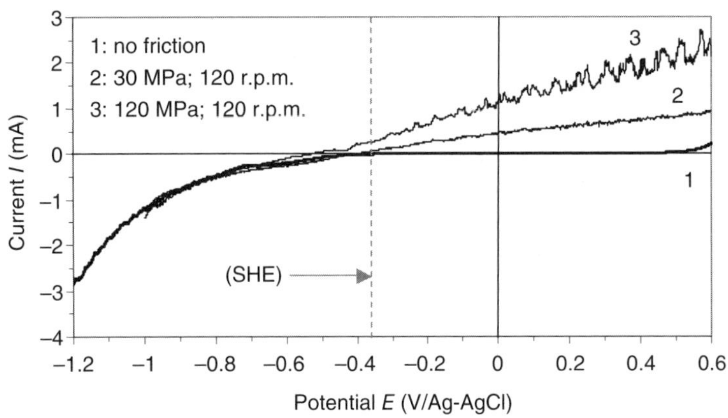

Fig. 11.6 Potentiokinetic polarization curves (corrected from ohmic drop due to the electrolyte resistance) of stellite 6 in water + B + Li, with no friction and under continuous sliding friction (30 and 120 MPa; 120 r.p.m.). Working surface: disc (4.9 cm^2). The potential of the standard hydrogen electrode (SHE) (−0.36 V/Ag-AgCl) is marked.

anodic end by the transpassive dissolution domain. The curves reveal that, in both electrolytes, there is no active dissolution peak. The magnitude of the anodic current on the passivation plateau is 5 μA in sulfuric acid and 10 μA in water + B + Li.

When friction is applied, the curve is not affected in the cathodic range. This indicates that the rates of the cathodic reactions are roughly the same on passive and depassivated areas. On the contrary, the curves in the anodic range are drastically changed. It is important to note that, at potential where passivation of the alloy occurs, most of the anodic current is flowing from the depassivated areas of the wear track. Out of the wear track, the alloy remains in a passive state and the corresponding current is probably very small, of the same order of magnitude as the passivation current measured when friction is not applied. Consequently, the anodic parts of the curves reveal the behavior of the alloy under friction in the wear track:

1 In H_2SO_4, an anodic active peak appears around –0.6 V/SSE. At higher potentials (between –0.45 up to +0.45 V/SSE) a current plateau is obtained with a characteristic current around 0.5 mA.
   - The existence of the active dissolution peak reveals that, around the potential value of –0.6 V/SSE, the passivation kinetics are slow and that, on bare alloy areas, the dissolution reaction prevails. In our previous study of tribocorrosion of stellite 6 in sulfuric acid [3], the existence of this narrow potential domain, where passivation is not effective enough to protect the alloy from dissolution, has provided an explanation of the gradual depassivation of the surface of stellite 6 specimens out of the wear track in the course of intermittent tribocorrosion tests.
   - The current plateau observed under friction between –0.45 and +0.45 V/SSE shows that, in this potential range, the kinetics of the oxidation reactions in the wear track are no longer controlled by the potential-dependent rate of the activation processes. The oxidation rate could be controlled by slow mass transfer processes, perhaps related to the passive film growth.

2 In the boric acid and lithia solution, the curves recorded under friction show that the passivation plateau disappears and a continuous increase of the anodic current with the potential is observed:
   - The above-mentioned limitation of the kinetics of oxidation reactions by mass transfer processes is not observed in this case. This could suggest that the film growth is very slow and that the film never becomes so thick that a limiting mass transfer can take place.

11.4.2 Potential jumps

Additional information can be obtained from potential jump experiments. For these tests, the following procedure was applied:

1 First, a cathodic galvanostatic polarization was applied to the specimen (disc; 4.9 cm^2) at a cathodic current density of 10 mA cm^{-2}. Hydrogen evolution, which occurs on the disc surface, induces cathodic reduction of the superficial oxide film. This method of 'cathodic depassivation' of the alloy is commonly used [4] in studies of the kinetics of the passive film growth on various stainless steels or passivating metals, by potential jump experiments. The conditions implemented in this work (10 mA cm^{-2}; 15 min) were determined from a preliminary study, in which the influence of various cathodic polarizations (current density; duration) on the electrochemical behavior of the alloy and on the response to potential jumps were tested.
2 The cathodic polarization was then switched off, and a potentiostatic polarization was instantaneously applied at –0.7 V/Ref (potential of the reference electrode). In both electrolytes, at this potential, hydrogen evolution is stopped, a zero current is measured, and it was shown by various measurements (impedance measurements) that the rate of oxidation of the alloy is very slow. Actually, in our experiments, this value of the potential is only applied for 2 s, which is the time necessary to remove the hydrogen bubbles from the surface with a soft brush. Remaining hydrogen bubbles could be oxidized during the anodic potential jump and induce an error in the determination of the passivation current.
3 Finally, a potential step is applied from –0.7 V/Ref to a more anodic value. The current transient is recorded for at least 3 min, at a frequency of 4 kHz.

In both electrolytes, potential jumps were applied to stellite 6 from the initial value of –0.7 V/Ref (SSE or Ag-AgCl) to final values of –0.4, –0.2, 0, +0.2, +0.4 V/Ref. It must be noted that, for both electrolytes, it was verified that the current response of the electrode was not limited by ohmic drop effect due to the electrolyte resistance.

Sulfuric acid solution

In Fig. 11.7, two current transients obtained in the sulfuric acid solution are presented as log–log scale. They are representative of the behavior of the alloy over the whole passivation domain from –0.4 to +0.4 V/SSE). The most significant result is that almost the same transient was obtained independent of the amplitude of the jump (provided that the final potential value was between –0.4 and +0.4 V/SSE). This surprising result is in fact consistent with the existence of the current plateau (Fig. 11.5) in the polarization curve of the alloy under friction. It can be interpreted in the same way: the passivation kinetics of a bare metal surface are independent of the applied anodic potential at potentials from –0.4 to +0.4 V/SSE.

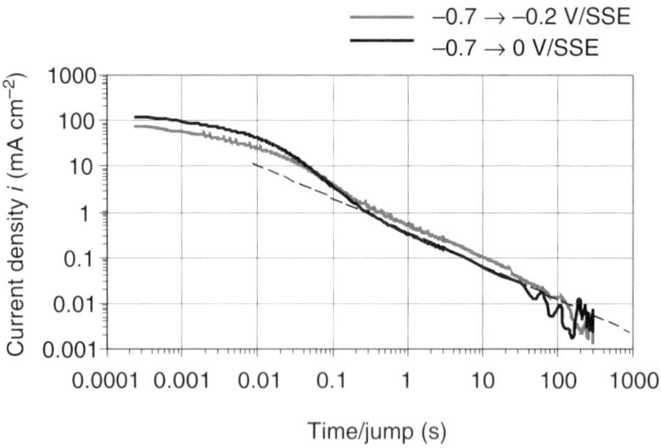

Fig. 11.7 Examples of current transients recorded on stellite 6 in 0.5 M H_2SO_4 after potential jumps (from –0.7 V/SSE to –0.2 and 0 V/SSE). SSE, Hg/H_2SO_4/saturated K_2SO_4 solution electrode.

The shape of the transients in Fig. 11.7 suggests that two stages might be distinguished:

1 From $t = 0$ to $t = 0.1$ s, the response of the working surface could be interpreted by two main processes: the charge of the double layer capacitance, and an electrochemical oxidation reaction leading to the adsorption of an intermediate acting as the 'precursor' from which the passive film will grow.
2 For $t > 0.1$ s, a linear decrease of log(i) versus log(t) is recorded with a slope of –0.7. Between 0.2 and 300 s, the current density variation is given by the following expression:

$$i(t) = i_0 \left[\frac{t}{t_0} \right]^{-p} \qquad [11.5]$$

with $p = 0.7$.

It is interesting to note that this expression has the same mathematical form as Eq. 11.2, which gives the variation of the current density during repassivation of a bare metal area under open circuit conditions. In fact, there is no clear justification for this similarity, because the transient described by Eq. 11.5 corresponds to potentiostatic conditions on a metal surface in a homogeneous state, whereas Eq. 11.5 corresponds to open circuit conditions, with a large evolution of the potential under galvanic coupling control conditions, on a heterogeneous surface. Anyway, the value of p is different from the value of n.

Such a linear decrease was also found on other materials and electrolytes in potential jump experiments (with other values of p) [2, 5, 6]. A

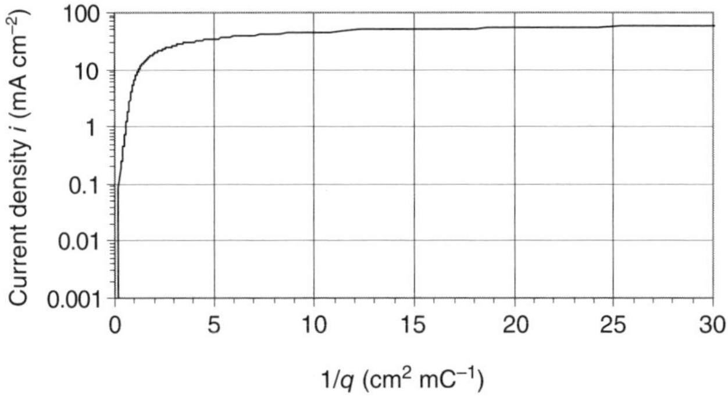

Fig. 11.8 Experimental relation between log[$i(t)$] and $1/q(t)$ obtained on stellite 6 during a test of potential jump from –0.7 to –0.2 V/SSE. SSE, Hg/H_2SO_4/saturated K_2SO_4 solution electrode.

few authors have interpreted this result as the consequence of the build-up of the passive film. Actually, the models of passive film growth, like the high field conduction model (HFCM) or its variants, cannot predict such a linear decrease of log(i) versus log(t) over more than three decades of current.

Nevertheless, the HFCM was considered further, because it could explain the results that suggest a limitation of the passivation kinetics by ionic mass transport through the film, and the invariability of the current response with the potential. One of the main predictions of the HFCM and of many of its variants is the linear relationship between log[$i(t)$] and $1/q(t)$, where $i(t)$ and $q(t)$ are respectively the current and total charge at time t:

$$\log[i(t)] = A + \frac{B}{q(t)}. \qquad [11.6]$$

An example of experimental relation is given in Fig. 11.8. We verified that, whatever the time scale is, a relation like Eq. 11.6 can never be clearly observed in our results. Consequently, it seems that the HFCM is not the appropriate model to develop a more detailed interpretation of our results and a realistic description of the behavior of stellite 6 in the sulfuric acid solution.

Water + B + Li

The current density transients obtained in the boric acid and lithia solution are represented in Fig. 11.9. Clear differences with the behavior in sulfuric acid are observed:

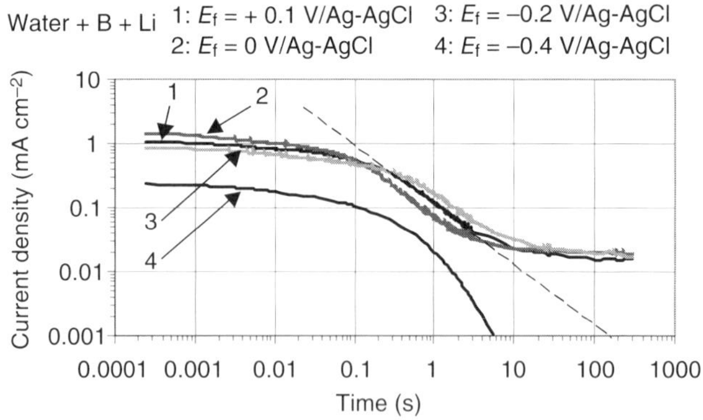

Fig. 11.9 Current transients recorded on stellite 6 in water + B + Li after potential jumps (from –0.7 V/Ag-AgCl to –0.4, –0.2, 0 and +0.1/Ag-AgCl). E_f: final potential.

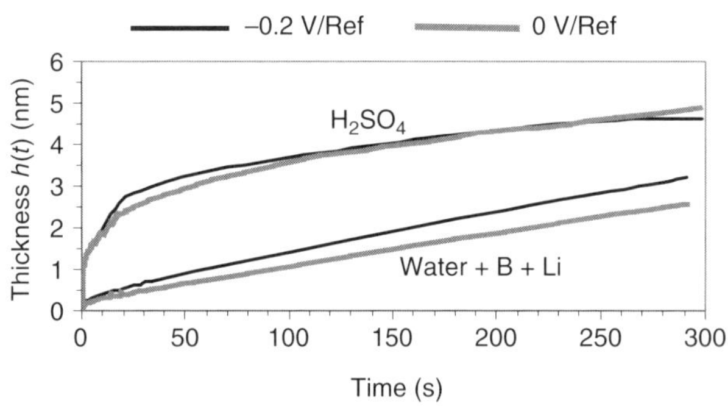

Fig. 11.10 Comparison of the evolution of the thickness of the passive film in H_2SO_4 and water + B + Li for two values of E_f (–0.2 V/Ref and 0 V/Ref).

1 The initial currents are much smaller (two orders of magnitude).
2 The linear decrease of log(i) versus log(t) is hardly visible over only one decade of current.
3 As in sulfuric acid, the current transients obtained for final potentials E_f between –0.2 and +0.4 V/Ref, are nearly identical, but the transients obtained for E_f < –0.2 V/Ag-AgCl are very different.

As in the case of sulfuric acid, it was not possible to obtain an experimental relationship log[$i(t)$] – 1/$q(t)$ in agreement with the theoretical Eq. 11.6 derived from the HFCM.

Additional information on the passivation mechanism and kinetics can also be derived if the variation of the current is analyzed in terms of film growth. In Fig. 11.10, the evolution of the thickness $h(t)$ of the film is represented for both electrolytes, for two jumps ($E_f = -0.2$ V/Ref and $E_f = 0$ V/Ref). To calculate these curves, it was first assumed that all of the anodic current during the transient was used for the growth of the film, and that no parallel anodic reaction (dissolution of components of the alloy for example) took place. The only reaction then considered is:

$$2Cr \rightarrow 2Cr^{3+} + 6e^-$$

$$2Cr^{3+} + 3H_2O \rightarrow Cr_2O_3 + 6H^+.$$

The current transient was integrated to obtain the charge variation and Faraday's law was used:

$$h(t) = \frac{1}{2}\frac{M}{zF\rho S}\int_0^t i(u)\mathrm{d}u \qquad [11.7]$$

where M is the molecular weight of Cr_2O_3 (= 152), z the valency of the chromium cations ($z = 3$), F the Faraday's number (96 500 C mol^{-1}), ρ the density of Cr_2O_3 ($\rho = 5.19$ g cm^{-3}) and S the working surface ($S = 4.91$ cm^2). It must be noticed that the values of $h(t)$ at the end of the transient (a few nanometers) are in agreement with the values reported in the literature for stainless steels at ambient temperature and pressure in an acidic environment. The increase of $h(t)$ is much faster in sulfuric acid, and identical evolutions are observed. This result was expected since the transients are also identical. On the contrary, the growth of the film in water + B + Li is much slower, and the curves do not merge. Surprisingly, the growth at 0 V/Ag-AgCl would be slower than at −0.2 V/Ag-AgCl.

If the values of $h(t)$ at short times (0.1 and 0.5 s = period in the tribocorrosion tests under continuous sliding friction) are compared, the following values are found:

0.5 M H_2SO_4:	−0.2 V/SSE	$h(0.1) = 0.60$ nm	$h(0.5) = 0.92$ nm
	0 V/SSE	$h(0.1) = 0.85$ nm	$h(0.5) = 1.10$ nm
Water + B + Li	−0.2 V/SSE	$h(0.1) = 0.03$ nm	$h(0.5) = 0.10$ nm
	0 V/SSE	$h(0.1) = 0.05$ nm	$h(0.5) = 0.12$ nm

This result could explain the differences in the polarization curves in H_2SO_4 and water + B + Li, and the existence of a plateau in the polarization curve in H_2SO_4:

1 One may consider that, in tribocorrosion tests in H_2SO_4, from 0.1 s after depassivation by the pin, a (thin) passive film has been built up. Therefore, the first step of adsorption of a 'precursor' (under kinetic control of charge transfer) seems to be a very fast reaction and after 0.1 s (or less) the current density on a small area of the wear track is related to the film growth which is independent of the potential, and limited by

ionic transfer in the film. This means that, under friction at 120 r.p.m., on most parts of the wear track, the growth of the film occurs. The current flowing from the wear track is then mainly due to this process and, as a result, independent of the potential.

2 On the contrary, we see that, in water + B + Li, after 0.1 s or even after 0.5 s, the film is not built up ($h \leq 0.12$ nm). Consequently, in tribocorrosion tests at 120 r.p.m., the first step of adsorption of the precursor still occurs on the whole wear track and the current is potential dependent (Fig. 11.6).

11.5 Conclusions

The comparison of the wear laws of stellite 6 found in this study, with the wear law determined on the worn gripper latch arms of pressurized water reactors, leads to the conclusion that similar tribocorrosion mechanisms are encountered in the PWR environment and in the laboratory in acid or neutral electrolytes at ambient pressure.

The mechanism involves a major influence of the passivation reaction responsible for the repair of the passive film at areas where it has been removed by friction. In the case of a periodic solicitation of the surface, the results can be explained in the following way: if the oxidation reaction is fast (in H_2SO_4 for example), the protective film is rapidly repaired after a friction step and its growth is also fast. As a result, at the end of the latency time, the following friction step removes a large amount of oxidized material, and the amount is larger if the latency time is longer, probably because the mechanical resistance of the film is low and most of the film is removed whatever its thickness may be. On the contrary, in a less oxidizing environment (such as water + B + Li), for the same duration of the latency period, the amount of passive film developed in the contact areas during the latency period is thinner, and the amount of oxidized material removed during the friction steps is smaller.

The kinetics of this process of film removal and repair determines the electrochemical component of wear, W_e. The mechanical component of wear W_m can be defined as the amount of material removed from the surface by friction on the bare metal. This component seems to have always a minor contribution in the total wear because the hardness and mechanical resistance of the alloy are very high, much higher than those of the passive film.

Beside passivation, the influence of a dissolution reaction can be considered, especially at bare metal areas in the wear track. In fact, the influence of this reaction has been pointed out, studied and explained in our previous paper on tribocorrosion of stellite 6 in 0.5 M H_2SO_4 [3]. In this study, its contribution to the wear in the tribocorrosion tests carried out in sulfuric acid has been eliminated. In the water + B + Li, and probably in the PWR environment, its contribution must also be negligible: the oxidizing properties of these electrolytes are weaker than those of H_2SO_4. The development of a protective passive film is then longer and more difficult but the dissolution is also much slower, leading to a negligible contribution of dissolution to wear.

The similarities between the mechanisms of tribocorrosion in the electrolytes and in the PWR environment permit us to envisage the development of tribocorrosion studies, easy to implement in laboratory conditions at low temperature and ambient pressure, with many powerful investigation techniques available for research work, to test for example the effect of changes in operating conditions (contact pressure, composition of the environment, etc.).

In addition, our results have shown that it was possible to rank the different environments (water + B + Li at 85°C, pressurized water, 0.5 M H_2SO_4 at 20°C) in relation to their oxidizing properties and aggressiveness towards stellite 6 in tribocorrosion conditions.

11.6 References

1. E. Lemaire, M. Le Calvar, *Wear*, **249**(5–6), 338–344, 2001.
2. F. P. Ford, P. L. Andresen, *Proc. Third Int. Symp. Environmental Degradation of Materials in Nuclear Power Systems – Water Reactors*, Traverse City, Michigan, USA, 30 August–3 September, pp. 789–800, 1987.
3. L. Benea, P. Ponthiaux, F. Wenger, J. Galland, D. Hertz, J. Y. Malo. *Wear*, **256**, 948–953, 2004.
4. P. Jemmely, S. Mischler and D. Landolt, *Wear*, **237**, 63–76, 2000.
5. E. A. Cho, Ch. K. Kim, J. S. Kim, H. S. Kwon. *Electrochim. Acta*, **45**, 1933–1942, 2000.
6. P. Andresen, F. P. Ford. *Corr. Sci.*, **38**(6), 1011–1016, 1996.

Index